WAVE PIPELINING: THEORY AND CMOS IMPLEMENTATION

THE KLUWER INTERNATIONAL SERIES IN ENGINEERING AND COMPUTER SCIENCE

VLSI, COMPUTER ARCHITECTURE AND DIGITAL SIGNAL PROCESSING

Consulting Editor

Jonathan Allen

Other books in the series:

CONNECTIONIST SPEECH RECOGNITION: A Hybrid Appoach, H. Bourlard, N. Morgan
ISBN: 0-7923-9396-1
BiCMOS TECHNOLOGY AND APPLICATIONS, SECOND EDITION, A.R. Alvarez
ISBN: 0-7923-9384-8
TECHNOLOGY CAD-COMPUTER SIMULATION OF IC PROCESSES AND DEVICES, R. Dutton, Z. Yu
ISBN: 0-7923-9379
VHDL '92, THE NEW FEATURES OF THE VHDL HARDWARE DESCRIPTION LANGUAGE, J. Bergé, A. Fonkoua, S. Maginot, J. Rouillard
ISBN: 0-7923-9356-2
APPLICATION DRIVEN SYNTHESIS, F. Catthoor, L. Svenson
ISBN:0-7923-9355-4
ALGORITHMS FOR SYNTHESIS AND TESTING OF ASYNCHRONOUS CIRCUITS, L. Lavagno, A. Sangiovanni-Vincentelli
ISBN: 0-7923-9364-3
HOT-CARRIER RELIABILITY OF MOS VLSI CIRCUITS, Y. Leblebici, S. Kang
ISBN: 0-7923-9352-X
MOTION ANALYSIS AND IMAGE SEQUENCE PROCESSING, M. I. Sezan, R. Lagendijk
ISBN: 0-7923-9329-5
HIGH-LEVEL SYNTHESIS FOR REAL-TIME DIGITAL SIGNAL PROCESSING: The Cathedral-II Silicon Compiler, J. Vanhoof, K. van Rompaey, I. Bolsens, G. Gossens, H. DeMan
ISBN: 0-7923-9313-9
SIGMA DELTA MODULATORS: Nonlinear Decoding Algorithms and Stability Analysis, S. Hein, A. Zakhor
ISBN: 0-7923-9309-0
LOGIC SYNTHESIS AND OPTIMIZATION, T. Sasao
ISBN: 0-7923-9308-2
ACOUSTICAL AND ENVIRONMENTAL ROBUSTNESS IN AUTOMATIC SPEECH RECOGNITION, A. Acero
ISBN: 0-7923-9284-1
DESIGN AUTOMATION FOR TIMING-DRIVEN LAYOUT SYNTHESIS, S. S. Sapatnekar, S. Kang
ISBN: 0-7923-9281-7
DIGITAL BiCMOS INTEGRATED CIRCUIT DESIGN, S. H. K. Embadi, A. Bellaouar, M. I. Elmasry
ISBN: 0-7923-9276-0
WAVELET THEORY AND ITS APPLICATIONS, R. K. Young
ISBN: 0-7923-9271-X
VHDL FOR SIMULATION, SYNTHESIS AND FORMAL PROOFS OF HARDWARE, J. Mermet
ISBN: 0-7923-9253-1
ELECTRONIC CAD FRAMEWORKS, T. J. Barnes, D. Harrison, A. R. Newton, R. L. Spickelmier
ISBN: 0-7923-9252-3
ANATOMY OF A SILICON COMPILER, R. W. Brodersen
ISBN: 0-7923-9249-3

WAVE PIPELINING: THEORY AND CMOS IMPLEMENTATION

C. Thomas GRAY
Wentai LIU
Ralph K. CAVIN, III
North Carolina State University
Raleigh, NC, USA

KLUWER ACADEMIC PUBLISHERS
Boston/London/Dordrecht

Distributors for North America:
Kluwer Academic Publishers
101 Philip Drive
Assinippi Park
Norwell, Massachusetts 02061 USA

Distributors for all other countries:
Kluwer Academic Publishers Group
Distribution Centre
Post Office Box 322
3300 AH Dordrecht, THE NETHERLANDS

Library of Congress Cataloging-in-Publication Data

Gray, C. Thomas, 1966 -
Wave pipelining : theory and CMOS implementation / C. Thomas Gray, Wentai Liu, Ralph K. Cavin, III.
p. cm. -- (The Kluwer International series in engineering and computer science. VLSI, computer architecture, and digital signal processing ; 248)
Includes bibliographical references and index.
ISBN 0-7923-9398-8
1. Metal oxide semiconductors, Complementary--Design and construction. 2. Digital integrated circuits--Design and construction. I. Liu, Wentai, 1948- . II. Cavin, Ralph K. 1939- . III. Title. IV. Series.
TK7871.99.M44G73 1993
621.39'732--dc20 93-27386
CIP

Printed on acid-free paper.

Printed in the United States of America

To Pamela, Bihjyu Lee, and Gwen

CONTENTS

LIST OF FIGURES

Chapter 3

Chapter 4

Chapter 5

Chapter 6

Chapter 7

LIST OF TABLES

PREFACE

The quest for higher performance digital systems for applications such as general purpose computing, signal/image processing, and telecommunications and an increasing cost consciousness have led to a major thrust for high speed VLSI systems implemented in inexpensive and widely available technologies such as CMOS. This monograph, based on the first author's doctoral dissertation, concentrates on the technique of wave pipelining as one method toward achieving this goal. The primary focus of this monograph is to provide a coherent presentation of the theory of wave pipelined operation of digital circuits and to discuss practical design techniques for the realization of wave pipelined circuits in the CMOS technology. Wave pipelining can be applied to a variety of circuits for increased performance. For example, many architectures that support systolic computation lend themselves to wave pipelined realization. Also, the wave pipeline design methodology emphasizes the role of controlled clock skew in extracting enhanced performance from circuits that are not deeply pipelined.

Wave pipelining (also known as maximal rate pipelining) is a timing methodology used in digital systems to increase the number of effective pipeline stages without increasing the number of physical registers in the pipeline. Using this technique, new data is applied to the inputs of a combinational logic block before the outputs due to previous inputs are available thus effectively pipelining the combinational logic and maximizing the utilization of the logic. Realization of practical systems using this technique requires accurate system-level and circuit-level timing analysis. At the system level, generalized timing constraints for the correct clocking of wave pipelined circuits are presented. Both single stage and multiple stage systems including feedback are considered. This work shows that the sizes of valid regions of operation are dependent on the clock period, the intentional clock skew, and the global clock latency. The minimum clock period is obtained by clock skew optimization formulated as a linear program. In addition to the generalized theory, important special cases are examined, and their relative performance limits are analyzed.

At the circuit level, since performance is determined by the maximum circuit delay difference for feasible signal paths, highly accurate estimates of both maximum and minimum delays are needed. It is shown that, for design methodologies such as wave pipelining where tight control is required on worst case circuit delays, traditional timing analysis based on gate delay models assuming single delay values for gates is not sufficient. For example, the delay of a two input CMOS NAND gate can vary by as much as a factor of two depending on whether one or both inputs undergo a transition. This implies that, for accurate detection of maximum and minimum overall delay, data dependent gate delay models must be used. To ascertain if it is possible for multiple simultaneous inputs to change at individual gate inputs, simultaneous sensitization of multiple paths is performed. While this problem is NP-complete for general circuits, efficient heuristics are demonstrated for exact analysis of small to medium sized circuits. The algorithm presented is implemented in a prototype timing analyzer *XTV* and results are given for a set of benchmark circuits.

Based on the theoretical results, practical techniques for the design of high speed wave pipelined circuits in CMOS are discussed. This includes a discussion of design approaches to manage noise, power variations, process variation, and the unique problems of data dependent delays in standard CMOS gates. Techniques to combat these problems are discussed and a CMOS logic family is presented that reduces data dependency of delay. The validity of these techniques and of the concept of wave pipelining in CMOS is shown for two circuits that have been implemented and tested. These are a 250 MHz wave pipelined 16-bit adder in MOSIS 2μm CMOS and a digital sampling circuit with 1 GBit/s bandwidth and 25 ps resolution in MOSIS 1.2μm CMOS.

The authors wish to thank the various groups that have provided financial support for this work. This support has included a Semiconductor Research Corporation Graduate Fellowship, a Design Automation Conference Fellowship, and a grant from the National Science Foundation MIP-9212346.

C. T. Gray
W. Liu
R. K. Cavin III

1

INTRODUCTION AND MOTIVATION

Increased performance of digital systems is critically important in applications ranging from general purpose computing to signal/image processing to telecommunications. In the design of digital integrated circuits, there are three primary approaches for achieving higher speed: technology, process, and design. First, to increase performance, the circuit can be implemented in a faster technology. For example, the performance of a CMOS circuit can be greatly increased by implementation in bipolar ECL or Gallium Arsenide (GaAs). However, with this performance, comes a significant increase in design and fabrication costs. Second, within a given technology, different processes can be used to obtain the desired performance. For example, performance is typically gained in CMOS technologies by shrinking minimum feature sizes. This increases the inherent speed of the devices and thus the performance of the system. Finally, once the technology and process are fixed, there are typically design options ranging from architecture to layout for obtaining higher performance. Techniques such as logic synthesis, retiming, transistor sizing, parallelism, and pipelining are routinely used to gain significant increases in system performance.

The concentration of this monograph is on obtaining optimal clocking of digital circuits by design using the technique of *wave pipelining*. That is, given a circuit in a specified technology, design a clock methodology such that maximal performance is achieved. In general, the CMOS technology is assumed although many of the formulations and results apply to other technologies as well.

1.1 WAVE PIPELINING

In performance driven integrated circuit design, the clocking methodology and the design of the clocking system are very important to overall system performance. Using conventional clocking methodologies, the minimal clock period is determined by the delay time of the latest possible transition at the circuit outputs of any combinational logic block. Thus, the designer expends major effort to reduce and optimize circuit paths that produce these worst case delays. Still, it is obvious that individual logic gates in the combinational logic blocks are idle for a vast majority of each clock cycle. For example, in modern CMOS processes, the gate switching time, from the time inputs arrive at a gate to the time at which the output of the gate is available, is typically less than 1 ns while the overall system clock period remains greater than 10 ns. This implies less than 10% logic utilization. That is, more than 90% of the clock cycle, individual logic gates are idle. This indicates a potential capability for much higher speed operation than is conventionally exploited. To improve utilization by minimizing this idle time and thus to allow for maximal rate operation of the circuit, the technique of *wave pipelining* is proposed.

To help illustrate this concept, a graphical representation [22] of the data flow through combinational logic is used. Figure 1.1 shows a *conventional* single stage system and its associated timing. As shown, this system consists of a single combinational logic block surrounded by edge triggered registers. At the beginning of each clock cycle, data is initiated into the logic block at the input register. Due to differences in circuit paths lengths and other factors, data delay through the combinational logic will vary. In the logic depth-timing diagram of Figure 1.1, the shaded regions, bounded by the maximal and minimal delays through the logic (t_{max} and t_{min}), depict the flow of data through the combinational logic and show time at which particular parts of the logic block are changing. The unshaded areas depict time at which the logic is stable. Thus, in the conventional system, the output register is clocked in the unshaded region after the latest data has arrived at the output allowing for the setup time of the register. At this time, the next input data is clocked into the logic at the input register. Therefore, the minimal clock period is bounded by the sum of the longest delay of the combinational logic and the propagation, setup, and hold times of the register along with any unintentional clock skew between input and output clocks.

The conventional method for decreasing the clock period and increasing the utilization of the logic is by inserting registers and *pipelining* the combinational logic block. For example, Figure 1.2 shows a system in which the combina-

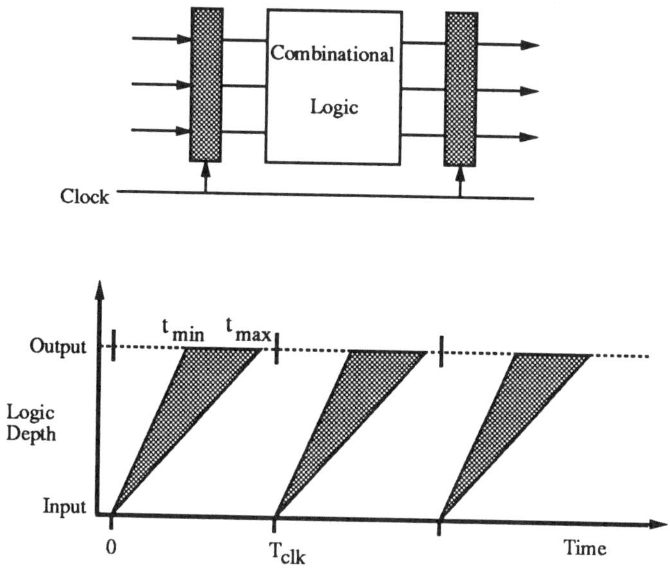

Figure 1.1 Combinational Logic with No Pipelining

tional logic block has been divided into two parts and a register inserted between. Now, each computation takes two clock cycles to propagate through the system, and a new computation is available at the output register each clock period. The timing diagram in Figure 1.2 demonstrates the constraints on the clock period of this system. In this case, since the maximal delay of each combinational logic block has been reduced, the overall system clock period can correspondingly be reduced. This process of dividing the circuit into increasingly smaller subblocks can theoretically be continued until each block consists of only one base level circuit element. However, there are two main factors that counteract the continued effectiveness of this approach. First, while the longest delay of each circuit block is reduced as more registers are inserted, the setup, hold, and propagation times of the registers are not. Therefore, since the clock period is bounded by sum of the logic block delay and the register parameters, the reduction in clock period is counteracted by the increasing dominance of the register parameters creating a situation of diminishing return. Second, the addition of more registers increases the demands on the clock distribution network. As the number of registers increases, the clock buffers must correspondingly be increased to drive the larger load. This adds to the overall power and area requirements of the circuit. Also, with an increased number of registers, the potential for increased clock skew exists. As mentioned above, this will also

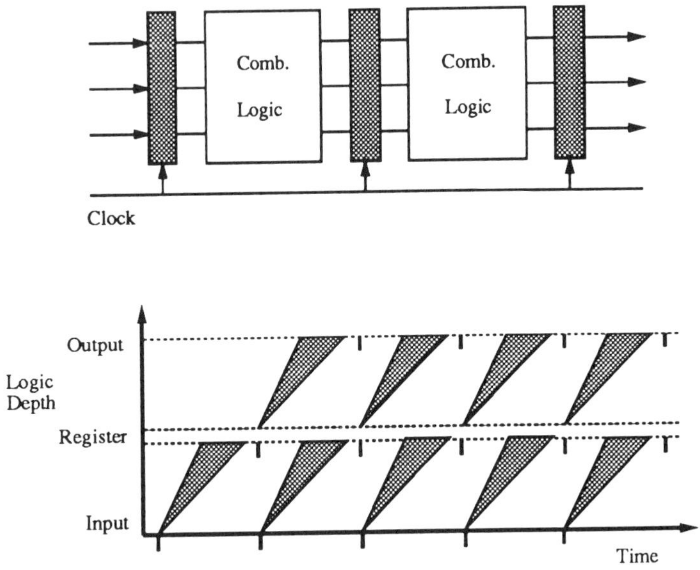

Figure 1.2 Combinational Logic with Conventional Pipelining

lengthen the clock period. Therefore, while conventional pipelining is certainly effective, its effectiveness is limited as more aggressive pipelining is attempted.

Revisiting Figure 1.1, the inputs to the output register are all stable for a significant portion of the clock period. This suggests that the combinational logic block itself is not operating at its maximal rate. In fact, the clock period can be reduced further as long as data from a particular clock cycle does not overwrite data from the previous clock cycle before the previous data has been clocked into the output register. In terms of the logic depth-timing diagram, the data uncertainty regions from adjacent clock periods must not overlap. Figure 1.3 shows a wave pipelined combinational logic block in which the clock period has been reduced as much as possible. The factor now limiting the clock period of this system is the *difference* in the maximal and minimal data delays (t_{max}-t_{min}) along with register setup, hold, and propagation times and unintentional clock skew (deviation from nominal).

While there are advantages and disadvantages for both aggressive conventional pipelining and wave pipelining, a generalized detailed comparison is difficult to make since many of the tradeoffs affecting the relative performance limits of these methods are highly application specific. For example, it is very diffi-

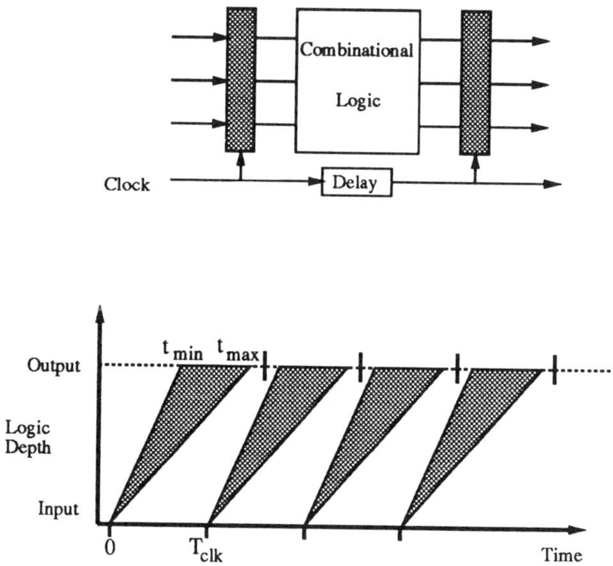

Figure 1.3 Combinational Logic with Wave Pipelining

cult, without specific knowledge of the technology and process to be used, to compare the relative difficulty of designing low skew, highly loaded clock distribution networks to the design of highly matched circuit delays or to compare the relative effects of process variation and environmental variations on absolute delays of conventionally pipelined systems to their effect on the difference of large delays in a wave pipelined system. However, it is clear from previous research and will be shown again in this research that wave pipelining can be useful in many applications in which conventional pipelining would be very difficult if not impossible. Therefore, since limits and constraints of conventional pipelining have been widely studied [51, 96, 17, 25], this research concentrates on the study of the limitations and constraints of wave pipelining. To potentially exploit the advantages of both approaches, it will also be shown how wave pipelining and conventional pipelining can be used together in multistage systems.

1.2 HISTORY

The idea of wave pipelining or maximal rate pipelining was first formalized by Cotten [18] in 1969 though it appears that similar ideas were used in the earlier designs of the IBM System/360 Model 91 floating point unit [1] and a maximal rate accumulator [62]. Recently, the concept has been a subject of renewed interest as technology and design techniques have allowed for effective implementation in integrated circuits and as clock periods are being pushed to physical technology limits. To date, work in the area of wave pipelining has included theoretical work on the timing constraints of wave pipelined systems [18, 51, 22, 30, 41, 83], work on techniques for path equalization [22, 109, 107, 49], and the design of systems using the techniques of wave pipelining [1, 78, 50, 107, 29, 24, 100, 73].

In the first of recent wave pipelining theoretical research, Ekroot[22] developed more fully a theory of the constraints for correct operation of wave pipelined systems and presented methods for inserting combinational logic delay units into wave pipelined systems to reduce the clock period by equalizing data delays. This optimization problem was formulated as a linear program. This work was instrumental in the resurgence of the wave pipelining methodology. Later, Wong [109, 108, 107, 110] proposed more practical methods for equalizing delays through combinational logic using ECL technology. This work consisted of rough tuning circuit delays by insertion of discrete delay units and then fine tuning delays by setting individual ECL gate drive currents. Using a linear programming formulation similar to Ekroot, Fishburn [26] presented a method for insertion of delay units along the clock paths to optimize system performance by clock skew optimization. The Fishburn formulation is a special case of the formulation developed in this monograph (see Section 2.6.4).

Concurrent with the work presented in this monograph, several groups of researchers have contributed to the development of theoretical constraints of wave pipelining. Joy and Ciesielski [40, 41, 39] have developed methods for placement of standard cells to minimize the clock period of wave pipelined systems and have also extended the Fishburn formulation to include internal node constraints and allow for multiple clock cycles between the initiating and capturing clocks for data in single stage systems. Klass and Mulder have presented new CMOS gate structures for low delay variation [50] and have developed additional methods for balancing circuit delays in CMOS circuits [49]. Srinivasan et. al. [93, 92] extended the Fishburn formulation to include stochastic delays and also developed a methodology for cycle time optimization under certain packaging constraints. Sakallah, et. al. [83, 12] have extended their previous

Table 1.1 Recent Research Relating to Wave Pipelining

Chiao-Tung U.	Currently designing wave pipelined multiplier [42] (CMOS)
Delft U.	Work on CMOS gates [50] and currently designing a 16x16 multiplier at Stanford U. (CMOS)
HP	Cache RAM in the HP Snake Workstation [52] (CMOS)
IBM	"Bubble pipelining" in RAM chip [13] (CMOS)
IBM	Crosspoint switch [90] (ECL)
NCSU	16-bit adder [24, 61], sampler [100, 102] and multiplier [73, 74] (CMOS)
Stanford U.	63-bit population counter [111] (ECL)
UC Berkeley	Single stage wave pipelining theory [54]
U. of Mass.	Wave pipelining in CMOS domino logic [59]
U. of Michigan	Multistage wave pipelining theory [83]

timing model [84] to allow for wave pipelined operation. Lien and Burleson [59] have developed constraints for correct clocking of wave pipelined CMOS domino circuits based on the Sakallah timing model. Lam, et. al. [54] have derived further relationships concerning intervals of correct clocking for a class of single stage wave pipelined circuits. These results by Lam are analogous to the special case results developed in Section 2.6.3.

In addition to the theoretical work, several recent circuit designs have used concepts of wave pipelining to increase system performance. Qi and Peisu [78] have designed a fast experimental computer in ECL consisting of an arithmetic/logic unit and a multiplication unit. Wong [107, 111], using his CAD tools developed for path equalization, designed and implemented a 63-bit population counter in ECL with a speed-up of 2.5. Using concepts similar to wave pipelining, Chappell et. al. [13] designed a 2 ns "bubble pipelined" RAM chip. Two other recent industry examples were published that use some degree of wave pipelining. These are the cache RAM access of the HP Snake workstation [52] and a crosspoint switch designed by IBM [90] for a wideband communication network. A summary of recent research relating to wave pipelining, including current on-going research, is given in Table 1.1.

1.3 DESIGNING WAVE PIPELINED CIRCUITS

There are three principle ingredients that are crucial to the design of working wave pipelined circuits:

- solid theoretical basis,
- computer aided design tools, and
- practical design techniques.

1.3.1 Theory

As shown in the intuitive discussion earlier, the main limiting constraint on the clock period (T_{clk}) of a wave pipelined circuit is the difference in the maximum and minimum circuits delays. This constraint is derived in Chapter 2 for the single stage system as:

$$T_{clk} \geq (t_{max} - t_{min}) + \text{setup/hold time} + \text{unintentional clock skew} \quad (1)$$

However, this is not the only constraint on the wave pipelined circuit clock period. In addition to insuring that there is sufficient time between data "waves" at the circuit output register, it must be assured that there is sufficient time between data waves at each internal circuit node. In other words, each circuit element must have enough time to respond to the present data before the next data arrives. Also, feedback in multistage systems can impose additional limits on the minimal clock period. All of these concerns point to the need for rigorous development of a theoretical foundation for correct clocking.

The theoretical work to date has examined many important aspects of timing constraints for wave pipelined circuits. However, as mentioned in the previous section, most work has concentrated on special or limiting cases. The purpose of this research is to present a unified theory of the constraints on circuit timing parameters for single stage and for multiple stage systems including the possibility of feedback.

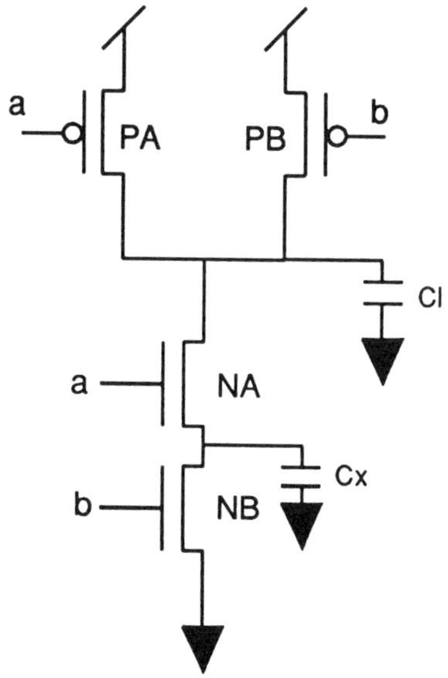

Figure 1.4 Static CMOS NAND Gate

1.3.2 CAD Tools

The core CAD capability for the design of high performance wave pipelined circuits is the ability to accurately extract delay information for combination circuits. While existing timing analysis techniques are sufficient for rough estimates of performance, accurate analysis of wave pipelined circuits requires more detailed timing analysis than is currently available. This is primarily due to the fact that wave pipelined circuits are designed to be nearly balanced and generally have tight timing requirements. This implies that very accurate delay models for base level circuit elements must be used.

To illustrate the level of accuracy needed, consider the delay of a standard static CMOS NAND gate as shown in Figure 1.4. The rising delay can vary by a factor of two depending on whether one input is falling or both inputs are falling. This implies that in order to determine exact delay information, it must be determined whether input nodes to individual circuit elements can switch near simultaneously.

This has an interesting implication on the method used for timing analysis of these circuits. The conventional method for determining circuit delays is extracting the longest and shortest graphical paths from the circuit and then checking to see if the logic structure of the circuit will allow a transition to propagate down the path under question. This is generally referred to as *path sensitization*. However, using accurate delay models, the delay of the path under question can potentially have many different delay values depending on the exact conditions of the remainder of the logic feeding into the path. This implies that, to determine if the worst case delay along the path can occur, all paths affecting the particular path under question must be analyzed.

1.3.3 Practical Techniques

The constraints for correct clocking of wave pipelined circuits also have important implications in the design of wave pipelined circuits in that the important design objective is now the balancing of maximum and minimum delay rather than the reduction of the maximum delay. This affects the selection of architecture for wave pipelined circuits, the selection of base level circuit elements, and the methods used for physical layout of wave pipelined circuits.

At the architecture level, it is important to choose an architecture which requires minimal delay balancing of slack paths. At the circuit level, it is important to choose gates structures whose delays do not vary widely depending on the gate input data. For example, the standard static NAND mentioned earlier (Figure 1.4) has a wide variation in delay and thus is not ideally suited for use in wave pipelined circuits. Finally, at the physical design level, it is important to use techniques to properly control interconnect loading and to minimize the effects of process and environmental variation and noise.

1.4 ORGANIZATION

This work addresses each of the three main concerns of the previous section and presents example integrated circuit designs in which the wave pipelining methodology has been used.

In Chapter 2, more generalized and detailed performance limits are derived for the single stage wave pipelined system and are shown to be consistent with previous special case results. Based on this formulation, several important new results are presented relating to performance limits of wave pipelined circuits. These results include the specification of distinct and disjoint regions of valid operation whose sizes are dependent on the clock period, intentional clock skew, and the global clock latency. Also, implications and motivations for the use of accurate delay models and exact timing analysis in the determination of combinational logic delays are given.

In Chapter 3, these results are extended to the general multi-stage wave pipelined system with feedback, and an analogous relationship between the multistage system and the single stage system in terms of performance limits is developed. This chapter concludes with an example of a three stage wave pipelined system that includes feedback.

With the degree of performance available using the wave pipelining theory of Chapters 2 and 3, accurate timing information for each combinational logic block is a necessity. The problem of extracting this information by timing analysis is examined in Chapters 4 and 5. It is shown, that particularly for CMOS logic gates, more detailed and accurate timing analysis is needed than is currently available. In particular, timing analysis for high performance CMOS circuits must consider the effects of data dependent gate delays.

Chapter 4 begins by discussing previous approaches to timing analysis and the need for accurate delay models that capture the effects of rising vs. falling delays, input terminal order dependent delays, and the difference in delays due to the relative arrive times of multiple input transitions. It is shown that the timing analysis problem under these conditions is NP-complete. In the second part of this chapter, a first straightforward algorithm is presented to solve the timing analysis problem. This approach is based on incremental exhaustive timing simulation of all possible input combinations. This algorithm is highly exponential in terms of run-time and space and thus is only feasible for small circuits, but it does present important notation and concepts that are used in a more efficient algorithm developed in Chapter 5.

The algorithm presented in Chapter 5 is based on storing possible transitions at logic nodes in an dependency graph and then searching this graph to eliminate transitions that cannot occur based on the logic constraints of the circuit. Results are presented for a prototype implementation of this algorithm (XTV), and it is shown to produce more accurate results that previous methods. Despite the exponential nature of the problem, runtimes are shown to be feasible for moderate size circuits. This chapter also presents extensions and enhancements to consider more accurate gate models and the effects of process and environmental variations.

Chapters 6 and 7 apply the results of the previous chapters to practical CMOS designs. Chapter 6 presents several practical considerations for the design of wave pipelined CMOS circuits. These considerations include choice of architecture, reduction of path delay variation, choice of base-level logic gates, the effects of process and environmental variations, and physical design.

Finally, Chapter 7 presents the design and implementation of two fabricated design examples. The first is a 250 MHz wave pipelined 16-bit adder in MOSIS $2\mu m$ CMOS and the second is a digital sampling circuit with 1 GBit/s bandwidth and 25 ps resolution in MOSIS $1.2\mu m$ CMOS. For both examples, an overview of the architecture is presented along with important design considerations and test results. These examples demonstrate the validity of the proposed theoretical and design methodologies and show the feasibility of wave pipelining using the CMOS technology.

2

CLOCK PERIOD CONSTRAINTS: SINGLE STAGE SYSTEMS

2.1 INTRODUCTION

The basis for any design methodology must be a firm theoretical foundation. A designer must be able to determine, given a proposed circuit, what the limits of performance are and how design parameters affect the performance. The next two chapters of this monograph investigate the theoretical performance limits of the wave pipelining design methodology and the effect of various circuit design parameters on the circuit clock period. It will be shown that, in contrast to conventional design techniques, performance is not limited by the longest absolute delay but is instead limited by differences in delay. In addition to presenting a rigorous theoretical formulation, it is a goal of this work to present intuitive insights into how various timing parameters interact.

The circuits on which the theory presented in these chapters is defined are represented at two levels. This is shown in Figure 2.1. The higher level view represents the circuit as a collection of interconnected combinational logic blocks and registers. Each of the combinational logic blocks is then represented at a lower level as a collection of interconnected circuit elements. Timing properties of each combinational logic block, each circuit element, and each register are assumed to be known.

Correct clocking of the system necessitates the coordination of clock and data signals at the inputs of each register such that valid relative timing is insured. Correct clocking also requires the coordination of various data edges within the combinational logic blocks. The presented theory describes the constraints on the interaction of circuit timing parameters under which correct clocking of the circuit is maintained and also describes the limits of the maximal rate of

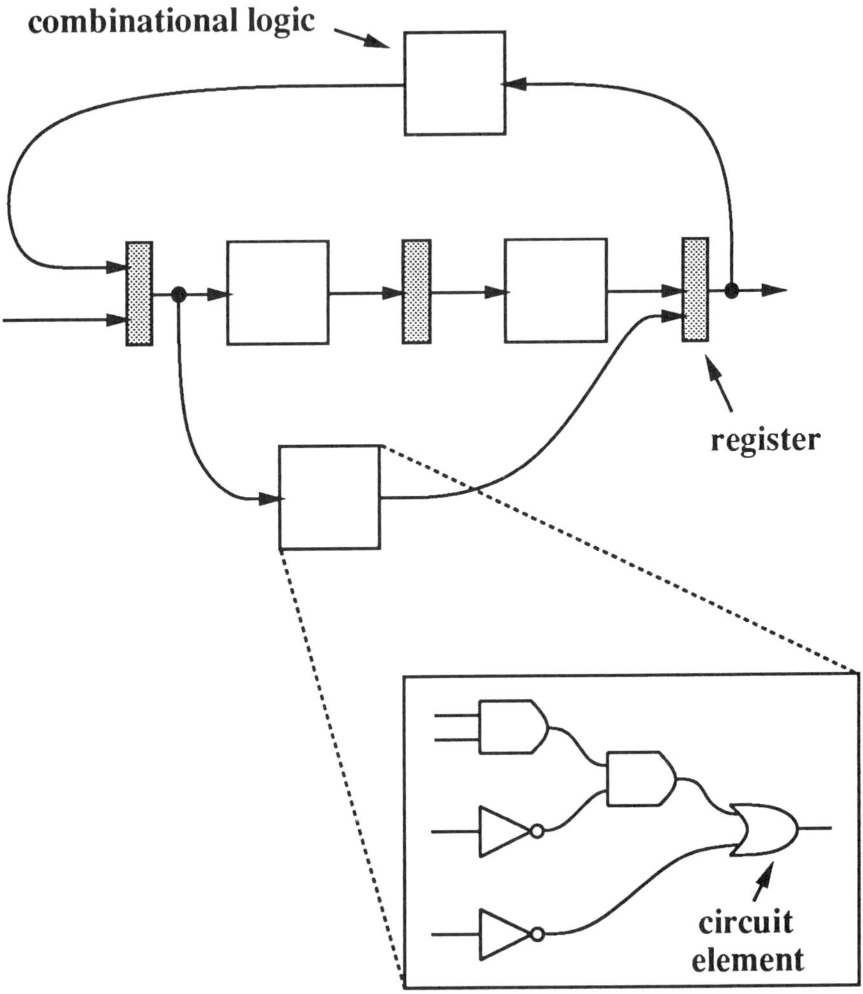

Figure 2.1 Circuit Representation

operation. For the designer, this relates to the maximal clock rate of the circuit and the conditions under which this rate can be achieved. Also discussed is the range on timing parameters for which the valid clocking is maintained. This relates to the reliability and manufacturability of the system.

With the ability to analyze the system, the next logical task is to optimize the system. That is, to determine what parameters could be changed to decrease the minimal clock period of the system and thus maximize performance. In the analysis theory, it will be shown that the performance of the system depends on the relative timing between multiple signals within the data path and also the relative timing between data and clock signal at the register inputs. Therefore opportunities for performance optimization exist within the data path and the clock generation circuitry.

These chapters concentrate on optimization by modifying the clock generation circuitry. This optimization method is based on the coordination of clock and data signals at the registers by clock skew optimization. It is assumed that the data path is fixed and that the only free circuit parameters are the clock skews to individual registers. Other work, as mentioned in Chapter 1, that relates to clock skew optimization can be found in [26, 41, 39, 83]. The main contribution of this work is the unification of many of the concepts presented in previous work along with the description of important new results and insights into the timing analysis and clock skew optimization of wave pipelined systems. Where appropriate, theoretical concepts are related to practical implementation to provide evidence for the feasibility of the wave pipelined methodology.

Several groups of researchers have also recently investigated various methods of optimizing the timing properties of the combinational logic blocks in the data path of wave pipelined systems by the modification of circuit delays [22, 108, 109, 40, 50]. This work has included insertion of discrete active delay elements, control of circuit element elements, and the insertion of passive RC delay in polysilicon routing. It is important to note that the optimization of combinational logic block timing can, in general, be done in addition to the clock skew optimization procedures presented in these chapters.

This chapter begins the theoretical study by investigating the constraints on performance of a simple single stage system. It is organized in seven sections. Section 2.2 presents a detailed view of the system model to be used. Using this model, Section 2.3 develops conditions and constraints for correct system clocking. Section 2.4 then gives a method for minimizing the clock period by clock skew optimization and Section 2.5 gives a discussion of the implication of the amount of global clock latency (k). Section 2.6 considers four impor-

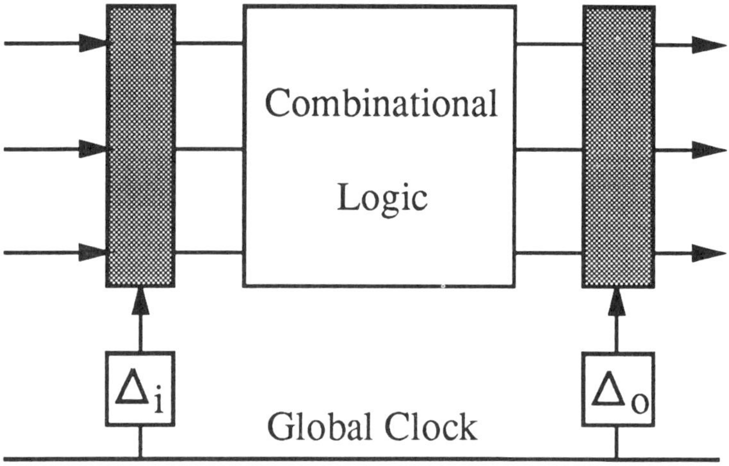

Figure 2.2 Single Stage Block Diagram

tant special cases of the general theory and analyzes their relative performance limits. Finally, Section 2.7 concludes the discussion of the single stage theory.

2.2 SYSTEM MODEL

2.2.1 Circuit Model

Figure 2.2 shows a block diagram of the single stage system model to be used in this paper. As shown, the data path consists of a single block of combinational logic surrounded by input and output edge triggered registers. It is assumed that all bits of the input register are synchronized with a single input clock and all bits of the output register are synchronized with a single output clock. These clocks are generated by the clock path which consists of delay units that produce delayed local versions of a global clock.

The combinational logic is described as a network of circuit elements. These circuit elements represent gates, collections of gates, or functional circuit blocks depending on the level of detail needed. Mathematically, the logic block is described as a graph with vertex set $\mathcal{N}$ representing the set of all circuit nodes (or wires) and edge set $\mathcal{C}$ representing the set of all connections between nodes

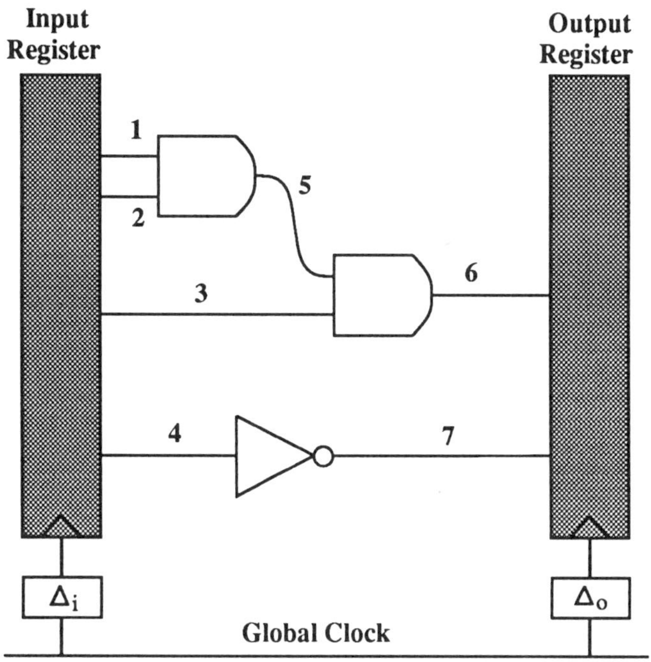

Figure 2.3 Single Stage Example Circuit Diagram

through circuit elements. For notational convenience, a set $\mathcal{G}$ is defined representing the set of all nodes that are circuit element output nodes and a set $\mathcal{O}$ is defined representing the set of all nodes that are outputs of the combinational logic block (that is, feeding into the output register). For example, in the combinational logic circuit shown in Figure 2.3, $\mathcal{N} = \{1, 2, 3, 4, 5, 6, 7\}$, $\mathcal{G} = \{5, 6, 7\}$, $\mathcal{O} = \{6, 7\}$, and $\mathcal{C} = \{(1, 5), (2, 5), (4, 7), (5, 6), (3, 6)\}$.

In summary, the following notation will be used to represent the single stage combinational logic block:

$\mathcal{N}$	set of all circuit nodes in logic block
$\mathcal{G}$	set of all gate output nodes in logic block
$\mathcal{O}$	set of all logic block output nodes
$\mathcal{C}$	set of all circuit connections

2.2.2 Timing Model

In addition to a physical interconnection of a circuit, timing properties of the circuit must be specified. The timing parameters can be divided into three categories: circuit timing parameters, clock timing parameters, and register timing parameters.

The *circuit* timing parameters of interest are:

t_{max}	maximum delay from input register to output register
t_{min}	minimum delay from input register to output register
$t_{max}(i)$	maximum delay from input register to internal node i
$t_{min}(i)$	minimum delay from input register to internal node i
$t_{stable}(i)$	minimum stable time for inputs to driving gate(s) of internal node i.

These parameters define the timing of data signal transitions. First, t_{max} and t_{min} refer to the maximum and minimum delay through the combinational logic between the input register and output register. These delay values are upper and lower bounds on the delay of the logic block over all expected process and environmental variation, over all circuit paths in the logic block, and for worst case input data patterns. Similarly, $t_{max}(i)$ and $t_{min}(i)$ $(\forall i \in \mathcal{G})$ represent upper and lower bounds on the delay from the input register to internal node i. The last parameter, $t_{stable}(i)$ $(\forall i \in \mathcal{G})$, represents the minimum amount of time that inputs to a circuit element or elements with output node i must be stable together in order for node i to respond. For example, in single level CMOS gates, this parameter will depend on the worst case input rise/fall times, output loading capacitance, transistor sizes, etc.

The *clock* timing parameters of interest are:

T_{clk}	clock period
Δ_i	intentional delay added to input clock
Δ_o	intentional delay added to output clock
Δ	Δ_o-Δ_i
Δt_e^n	earliest possible clock skew at input register
Δt_l^n	latest possible clock skew at input register.
Δt_e^o	earliest possible clock skew at output register
Δt_l^o	latest possible clock skew at output register

These parameters define the timing of clock signal transitions. T_{clk} represents the period of the clock. Δ represents the effective amount of intentional skew between the local input clock and the local output clock. This is implemented by the delay units shown in Figure 2.2. The last four parameters, Δt_e^n, Δt_l^n, Δt_e^o, and Δt_l^o, represent various unintentional skew at the input and output registers. These skews are defined relative to a nominal arrival time and are due to variations in intentional delay units, variations in physical clock interconnect, clock jitter, etc.

The *register* timing parameters of interest are:

t_d	delay time of register
t_{setup}	set up time of register
t_{hold}	hold time of register.

These parameters define the relationship between clock and data edges as determined by the register structure being used. First, t_d defines the amount of delay between the clock edge at the register and the output data transition. t_{setup} defines the amount of time that input data must be stable *before* the clock edge arrives and t_{hold} defines the amount of time that input data must be stable *after* the clock edge arrives.

2.3 CONSTRAINTS FOR CORRECT CLOCKING

2.3.1 Register Constraints

In order to achieve proper capturing of data at the output register using the model described above, relatively stringent double sided timing requirements must be met between clock and data edges at the inputs to the output register. The clock period must be such that the output data is latched *after* the latest data has arrived at the outputs and *before* the earliest data from the next clock period arrives.

As shown in the logic depth-timing diagram of Figure 2.4, output data will be latched by the output clock at some time (t) relative to the global clock. This output clock, as described above, will be derived from the global clock delayed by a physical delay unit of length Δ_o. Given a physical delay Δ_o and a target

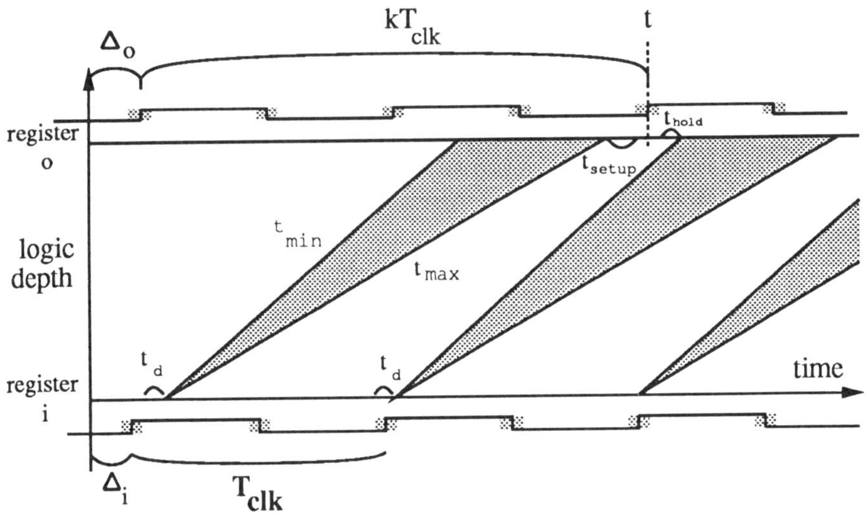

Figure 2.4 Constraints at Registers

time t, there are several options for achieving this skew. First, Δ_o could be equal to t. This would achieve the proper skew and allow for proper operation. Alternately, the cyclic nature of the clock could be relied upon to effect part of the shift. The portion of t achieved by this mechanism is kT_{clk} where k is an integer referring to the number of clock cycles. The parameter k will be referred to as the *global clock latency* and will be described in more detail in Section 2.5. Thus t can be expressed as the sum of the amount of clock delay derived from the cyclic nature of the clock (kT_{clk}) and the amount of delay produced by the physical delay unit (Δ_o). That is,

$$t = kT_{clk} + \Delta_o.$$

Therefore, it is obvious that, for correct clocking, the output data must be sampled at a time (t) at which the data is stable (that is, t must fall in a non-shaded region in Figure 2.4). This implies that, to ensure the latest data is latched correctly, the following constraint must be true:

$$kT_{clk} + \Delta_o - \Delta t_e^o \geq \Delta_i + \Delta t_l^n + t_d + t_{max} + t_{setup} \quad (1)$$

where $kT_{clk} + \Delta_o = t$ as described above.

And, to ensure the earliest data is latched correctly, the following constraint must be also true:

$$\Delta_i + T_{clk} + t_d + t_{min} - t_{hold} - \Delta t_e^n \geq kT_{clk} + \Delta_o + \Delta t_l^o \quad (2)$$

Now, by combining constraints (1) and (2) and rearranging terms, the following condition on the number of waves is necessary for proper latching:

$$T_{clk} + t_{min} - t_{hold} - \Delta t_e^n - \Delta t_l^o \geq kT_{clk} + \Delta - t_d \geq t_{max} + t_{setup} + \Delta t_l^n + \Delta t_e^o \quad (3)$$

By transitivity, constraint (3) then implies that constraint (4) must be satisfied.

$$T_{clk} \geq (t_{max} - t_{min}) + t_{setup} + t_{hold} + \Delta t_e^n + \Delta t_e^o + \Delta t_l^n + \Delta t_l^o \quad (4)$$

Thus, the maximal clock rate is bounded by the difference of maximal and minimum delays between input and output registers. It is also evident that setup and hold times along with unintentional clock skew directly increase the possible clock period.

2.3.2 Internal Node Constraints

Physical limits on switching delay of individual gates impose additional constraints on the minimal clock period. Constraint (4) indicates that the clock period is bounded only by setup/hold time and clock skew when input to output logic paths are equalized ($t_{max} = t_{min}$). However, circuit factors, such as non-zero rising/falling time, transistor sizing, capacitive loading, signal dispersion, and circuit technology, certainly can affect the clock speed. Accordingly, additional constraints due to physical properties of *internal nodes* must be considered in addition to register constraints such that fundamental circuit limitations are not exceeded and so that data waves from separate clock periods do not intermix at any of the internal nodes. In other words, the next earliest possible wave at any particular node i should not arrive until the latest possible wave has propagated through the node. This will be insured if all inputs to each circuit element are stable long enough for the node to respond before the next data arrives. With the amount of stable time needed at the inputs of the circuit element driving node i represented by $t_{stable}(i)$ as defined in Section 2.2.2, these constraints at all internal nodes are expressed by constraint (5).

$$T_{clk} + \min_{(i,j)\in\mathcal{C}}(t_{min}(i)) - \Delta t_e^n \geq \max_{(i,j)\in\mathcal{C}}(t_{max}(i)) + \Delta t_l^n + t_{stable}(j), \qquad \forall j \in \mathcal{G} \tag{5}$$

In order to compare the internal node constraints with the register constraints, note that constraint (5) can be rewritten as:

$$T_{clk} \geq \max_{(i,j)\in\mathcal{C}}(t_{max}(i)) - \min_{(i,j)\in\mathcal{C}}(t_{min}(i)) + t_{stable}(j) + \Delta t_e^n + \Delta t_l^n \qquad \forall j \in \mathcal{G} \tag{6}$$

Note that, if $t_{max}(i)$, $t_{min}(i)$, and $t_{stable}(i)$ are constant for all i in $\mathcal{G}$, the set of constraints of (6) can be written as one constraint (7).

$$T_{clk} \geq \max_{\forall j\in\mathcal{G}}(\max_{(i,j)\in\mathcal{C}}(t_{max}(i)) - \min_{(i,j)\in\mathcal{C}}(t_{min}(i)) + t_{stable}(j)) + \Delta t_e^n + \Delta t_l^n \tag{7}$$

Given the similar form of this constraint (7) and the constraint (4) derived from the register constraints, the question naturally arises as to whether the internal node constraint is needed or if it is in fact implied by the register constraints. Conditions under which the constraint is not needed are expressed in the following lemma and theorem.

Lemma 2.1 *The internal node constraint (7) is not active if:*

$$t_{max} - t_{min} + t_{setup} + t_{hold} + \Delta t_e^o + \Delta t_l^o \geq \max_{\forall j\in\mathcal{G}}(\max_{(i,j)\in\mathcal{C}}(t_{max}(i)) - \min_{(i,j)\in\mathcal{C}}(t_{min}(i)) + t_{stable}(j)) \tag{8}$$

Proof: Defining

$$t_{internal} = \max_{\forall j\in\mathcal{G}}(\max_{(i,j)\in\mathcal{C}}(t_{max}(i)) - \min_{(i,j)\in\mathcal{C}}(t_{min}(i)) + t_{stable}(j)) + \Delta t_e^n + \Delta t_l^n$$

and

$$t_{register} = t_{max} - t_{min} + t_{setup} + t_{hold} + \Delta t_e^o + \Delta t_l^o + \Delta t_e^n + \Delta t_l^n$$

if $t_{register} \geq t_{internal}$ as stated in the lemma and $T_{clk} \geq t_{register}$ as shown in constraint (4), by transitivity, this implies that $T_{clk} \geq t_{internal}$ which is the constraint (7). Thus, constraint (7) is not an active constraint under these conditions. □

Note that under certain conditions the inequality of Lemma (2.1) can be guaranteed to be satisfied.

Theorem 2.1 *The internal node constraint (7) is not active if both of the following conditions are satisfied:*

(i) $t_{setup} + t_{hold} \geq t_{stable}(i)\, \forall i \in \mathcal{G}$

(ii) $t_{max}(j) - t_{min}(j) \geq \max_{(i,j)\in\mathcal{C}}(t_{max}(i)) - \min_{(i,j)\in\mathcal{C}}(t_{min}(i))\, \forall j \in \mathcal{G}$

Proof: By the definition of t_{max} and t_{min},

$$t_{max} \equiv \max_{\forall i \in \mathcal{O}}(t_{max}(i)) \geq t_{max}(j) \forall j \in \mathcal{O} \tag{9}$$

and

$$t_{min} \equiv \min_{\forall i \in \mathcal{O}}(t_{min}(i)) \leq t_{min}(j) \forall j \in \mathcal{O}. \tag{10}$$

Now, by adding inequalities (9) and (10), the following inequality is implied.

$$t_{max} - t_{min} \geq t_{max}(j) - t_{min}(j) \forall j \in \mathcal{O} \tag{11}$$

By condition *(ii)*,

$$t_{max}(j) - t_{min}(j) \geq \max_{(i,j)\in\mathcal{C}}(t_{max}(i)) - \min_{(i,j)\in\mathcal{C}}(t_{min}(i)) \forall j \in \mathcal{O}. \tag{12}$$

And, since

$$\max_{(i,j)\in\mathcal{C}}(t_{max}(i)) - \min_{(i,j)\in\mathcal{C}}(t_{min}(i)) \geq t_{max}(k) - t_{min}(k) \forall (k,j) \in \mathcal{C}\, \forall j \in \mathcal{G} \tag{13}$$

by transitivity

$$t_{max} - t_{min} \geq t_{max}(j) - t_{min}(j) \forall j \in \mathcal{G}. \tag{14}$$

By condition (*ii*)

$$t_{max}(j) - t_{min}(j) \geq \max_{(i,j)\in\mathcal{C}}(t_{max}(i)) - \min_{(i,j)\in\mathcal{C}}(t_{min}(i))\ \forall j \in \mathcal{G} \tag{15}$$

so therefore

$$t_{max} - t_{min} \geq \max_{\forall j\in\mathcal{G}}(\max_{(i,j)\in\mathcal{C}}(t_{max}(i)) - \min_{(i,j)\in\mathcal{C}}(t_{min}(i))). \tag{16}$$

Notice that condition (*i*) of Theorem 2.1 can be rewritten as

$$t_{setup} + t_{hold} \geq \max_{\forall i\in\mathcal{G}}(t_{stable}(i)). \tag{17}$$

Adding inequalities (16) and (17) gives condition

$$\begin{gathered} t_{max} - t_{min} + t_{setup} + t_{hold} \geq \\ \max_{\forall j\in\mathcal{G}}(\max_{(i,j)\in\mathcal{C}}(t_{max}(i)) - \min_{(i,j)\in\mathcal{C}}(t_{min}(i))) + \max_{\forall j\in\mathcal{G}}(t_{stable}(j)) \end{gathered} \tag{18}$$

which implies the following condition.

$$\begin{gathered} t_{max} - t_{min} + t_{setup} + t_{hold} + \Delta t_e^o + \Delta t_l^o \geq \\ \max_{\forall j\in\mathcal{G}}(\max_{(i,j)\in\mathcal{C}}(t_{max}(i)) - \min_{(i,j)\in\mathcal{C}}(t_{min}(i)) + t_{stable}(j)). \end{gathered} \tag{19}$$

By Lemma 2.1, the theorem is proved. □

Condition (*ii*) of Theorem 2.1 is satisfied when the delay difference at each node is monotonically increasing from inputs to outputs as shown in Figure 2.5. That is, at each circuit element, the difference in delay is greater at the circuit element output than at any of its inputs. This implies that the smallest stable region between successive "waves" will be at the inputs of the output registers. Condition (*i*) of Theorem 2.1 is satisfied when the register performance is equal to or slower than the circuit element performance. This implies that the required stable region will be greatest at the inputs of the output register (due to the required setup and hold times). Therefore since the stable region is always larger at internal nodes than at output nodes and since the required stable time is greatest at the outputs, then, if timing requirements are met at the output register ($t_{setup}+t_{hold}$), the timing requirements at internal nodes ($t_{stable}(i)$) will be automatically met.

Whether or not the condition (*i*) is satisfied simply depends on the relative timing characteristics of the output register and the internal logic circuitry.

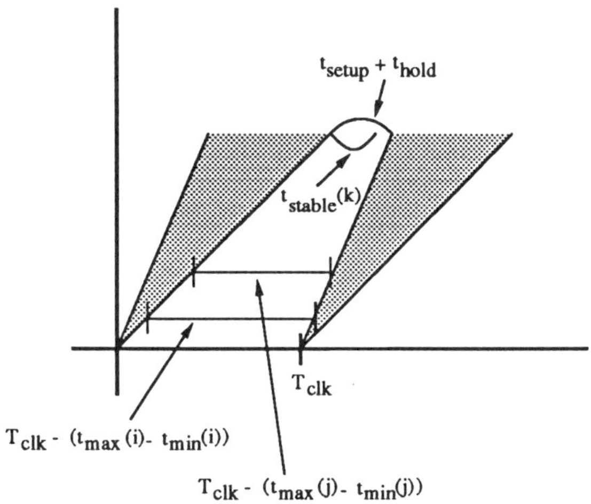

Figure 2.5 Internal Node Timing

Whether or not condition (*ii*) is true depends on the method used to calculate delays in circuit paths and the models used for the gate delays. In general, as shown in the following examples, this condition cannot be guaranteed to be satisfied if delay computation methods using accurate delay models and eliminating false delay paths are used.

First, consider the effect of different delay models. Figure 2.6 shows a simple circuit consisting of a single NAND gate and a delay unit. In Figure 2.6.a, the NAND gate is assumed to have a single delay independent of the gate input. The maximum and minimum values of this delay are shown to be 2 and 1 time units respectively. As can be seen from the space-time diagram, condition (*ii*) of Theorem 2.1 is satisfied in this case. In Figure 2.6.b, a more realistic model is used in which the NAND gate is assumed to have a delay dependent on the gate input[87]. In this case, condition (*ii*) of Theorem 2.1 is not satisfied.

The implications for false path elimination are demonstrated in the following example taken from [76] and shown in Figure 2.7. The circuit consists of two multiplexors, four delay elements, and two inverters between the input and output registers. The numbers inside each circuit element represents the delay of that element and $t_{setup} = t_{hold} = t_{stable}(i)$ $(\forall i) = 2$. The timing diagram in Figure 2.7 graphically depicts the timing constraints at the registers and at two internal nodes (the outputs of the two multiplexors) based on a graph level timing analysis. For example, the longest delay from input register to

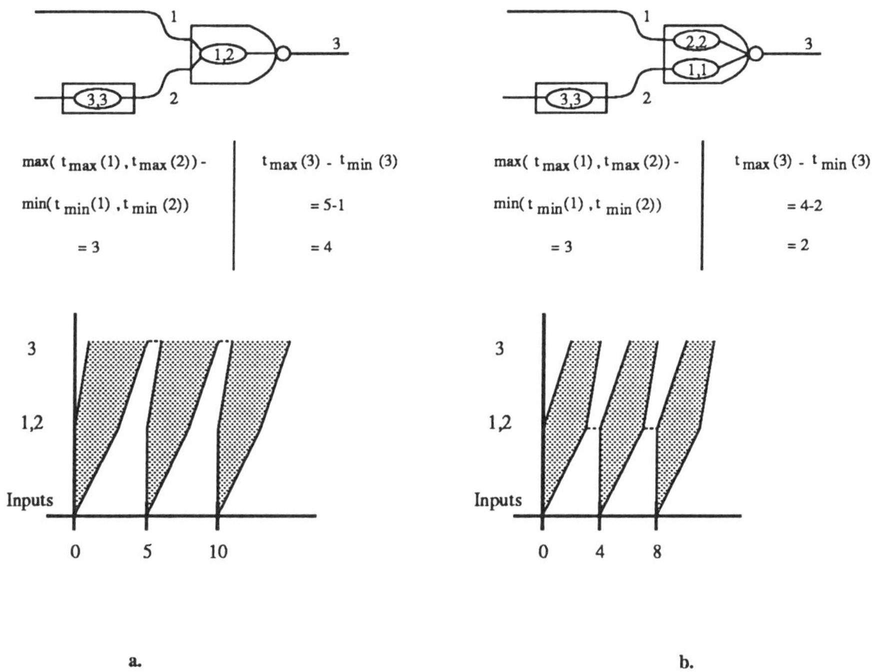

Figure 2.6 Gate Delay Model Example

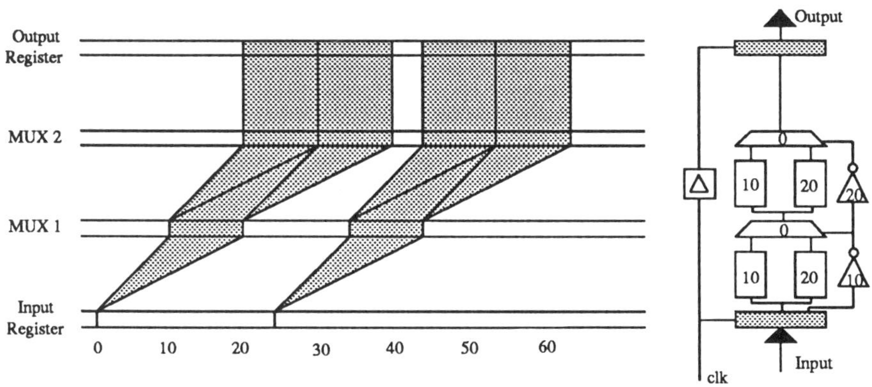

Figure 2.7 Example Not Considering False Paths ($t_{setup}= t_{hold}= t_{stable}(i) = 2$)

output register (t_{max}) is 20+20=40 and the shortest delay (t_{min}) is 10+10=20. It is evident that condition (*ii*) of Theorem 2.1 is satisfied under this delay computation method, and the figure shows that the limiting constraint is indeed at the output. The minimum clock period is 24 (t_{max}-t_{min}+t_{setup}+t_{hold}).

However, further investigation of the circuit reveals that, in fact, the path through the two elements of length 20 can never be selected due to the multiplexor control. Therefore, this path is a false path. In fact, no matter which path is selected the only true paths are of length 30 ($t_{max}= t_{min}= 30$). Thus, if only the constraints at the output are considered, the conclusion would be that the minimum clock period is bounded only by setup and hold times along with unintentional clock skew. This would be an erroneous conclusion, however, since this would cause data overlap at the output of the first multiplexor. The true minimum clock period is shown in Figure 2.8 considering the constraint at the internal node. Note that the minimum clock period is now 12 ($t_{max}(i)$-$t_{min}(i)$+$t_{stable}(i)$). Therefore the clock period is reduced when actual circuit delays are used, but now the designer must consider the constraints at all internal nodes since condition (*ii*) is not satisfied.

In conclusion, unless the designer can be sure that internal node constraints are not needed, for example by using Theorem 2.1, the internal node constraints must be considered. This will depend on the relative speeds of the register and logic circuitry and the method and models used to compute circuit delays. In general, for operation at the true wave pipelining performance limits, accurate timing analysis must be done and the internal nodes must be considered. Fur-

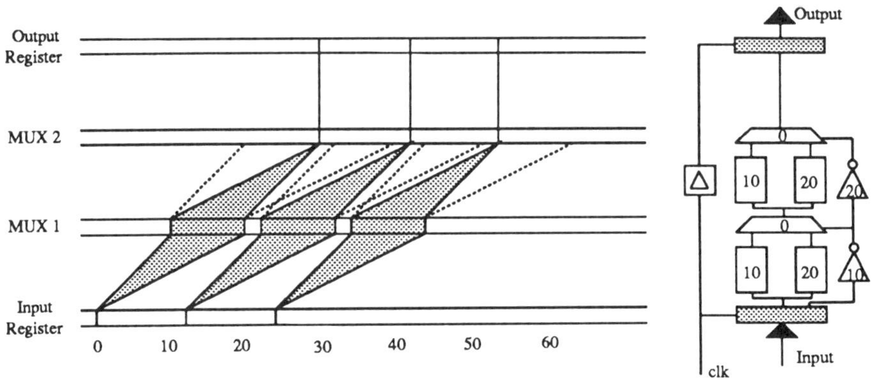

Figure 2.8 Example Considering False Paths ($t_{setup} = t_{hold} = t_{stable}(i) = 2$)

ther comments will be made on the calculation of circuit delay in Chapters 4 and 5.

2.3.3 Constraint Space Diagram

Figure 2.9 graphically illustrates the register and internal node timing constraints presented in the previous sections. For clarity, setup time, hold time, register delay, and unintentional clock skew have been set to zero. Also, the lower bound on clock period due to internal node constraints is defined to be $t_{internal}$:

$$t_{internal} = \max_{\forall j \in \mathcal{G}}\left(\max_{(i,j)\in \mathcal{C}}(t_{max}(i)) - \min_{(i,j)\in \mathcal{C}}(t_{min}(i)) + t_{stable}(j)\right) + \Delta t_e^n + \Delta t_l^n \quad (20)$$

The feasible solution space for each value of k is a convex region (shaded in Figure 2.9) bounded by $T_{clk} + t_{min} = kT_{clk} + \Delta$, $kT_{clk} + \Delta = t_{max}$, and $T_{clk} = t_{internal}$(as shown the internal node constraint is not active). For fixed values of k, distinct feasible regions of operation are available. For example, when $k = 2$, the feasible region is bounded by $T_{clk} + t_{min} = 2T_{clk} + \Delta$ and $2T_{clk} + \Delta = t_{max}$. Note that as k is increased, the feasible region is reduced.

This figure shows an example in which the clock period is bounded below by register constraints (t_{max}-t_{min}). From the figure is easy to see that this is indeed true for all values of k and that the feasible region for a particular value

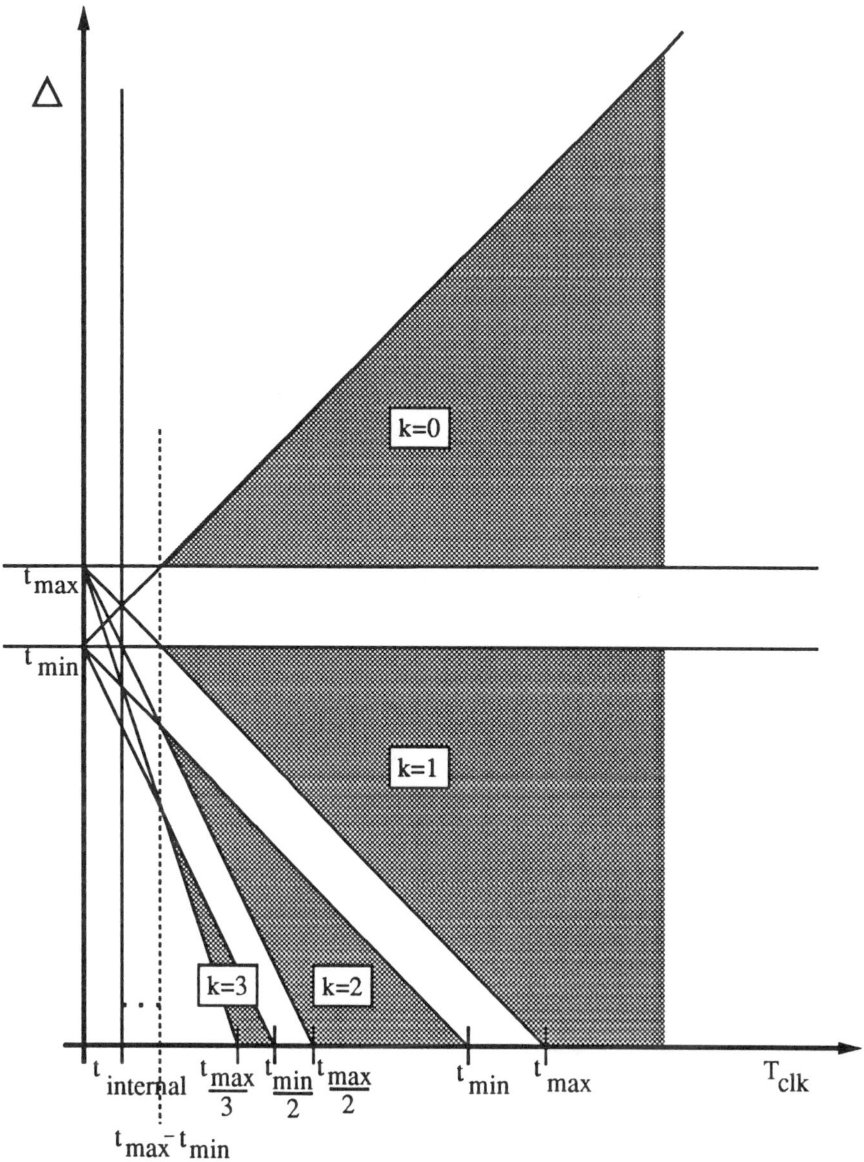

Figure 2.9 Timing Constraint Space Diagram

of k is shrunk as t_{max}- t_{min} is increased. This emphasizes the necessity of delay balancing for maximal clock frequency. Figure 2.10 demonstrates the case where the internal node constraint is the limiting factor at time $t_{internal}$.

2.4 MINIMIZING THE CLOCK PERIOD

Given the constraints identified in the previous section, the clock period (T_{clk}) can now be minimized. Given variable T_{clk} and Δ and all other parameters fixed, this can be formulated as a linear programming problem:

Min: T_{clk}

Subject to:

$$kT_{clk} + \Delta - \Delta t_e^o \geq \Delta t_l^n + t_d + t_{max} + t_{setup}$$

$$T_{clk} + t_d + t_{min} - t_{hold} - \Delta t_e^n \geq kT_{clk} + \Delta + \Delta t_l^o$$

$$T_{clk} + \min_{(i,j)\in C}(t_{min}(i)) - \Delta t_e^n \geq \max_{(i,j)\in C}(t_{max}(i)) + \Delta t_l^n + t_{stable}(j), \quad \forall j \in \mathcal{G}$$

Theorem 2.2 *The minimum clock period (T^*_{clk}) of the single stage system is*

$$\max(t_{internal}, t_{max} - t_{min} + t_{setup} + t_{hold} + \Delta t_e^n + \Delta t_e^o + \Delta t_l^n + \Delta t_l^o). \quad (21)$$

Proof: Since this problem is formulated as a linear program, some vertex of the feasible region will give the optimal solution. Thus, it is only necessary to consider the intersection of feasible region boundaries. Also, as shown in Section 2.3.2, the last constraint may or may not be an active constraint. Thus, there are two cases for the optimal solution.

Case 1: Internal Node Constraint Active
If the last constraint is active, the feasible region is obviously bounded at

$$T_{clk} = t_{internal}$$

since this constraint does not depend on Δ or k.

Case 2: Internal Node Constraint Not Active
If the last constraint is not active, the feasible region will be defined by the

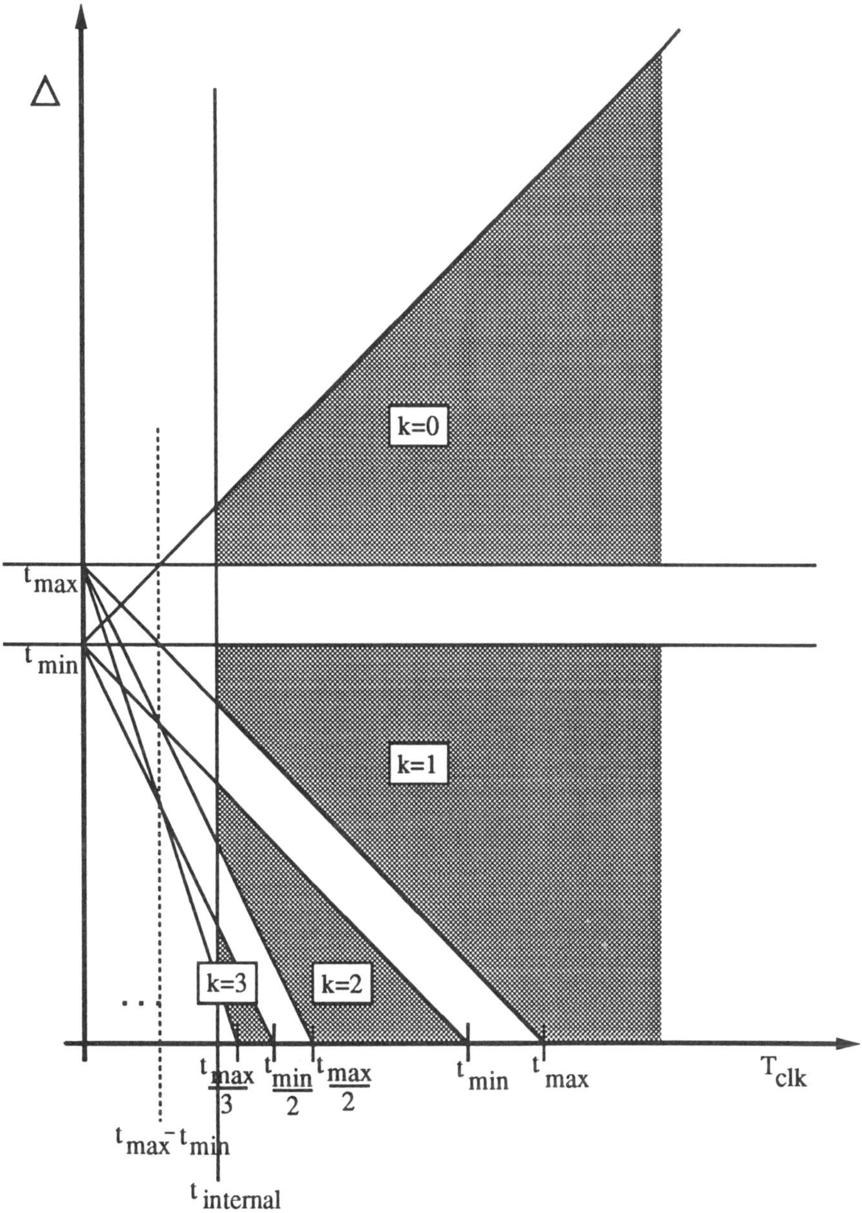

Figure 2.10 Timing Constraint Space Diagram with Internal Node Constraint

first two constraints. Thus the optimal solution will be at the intersection of the boundaries of the feasible regions defined by these two constraints. That is, the optimal solution will be at the intersection of

$$kT_{clk} + \Delta - \Delta t_e^o = \Delta t_l^n + t_d + t_{max} + t_{setup}$$

and

$$T_{clk} + t_d + t_{min} - t_{hold} - \Delta t_e^n = kT_{clk} + \Delta + \Delta t_l^o.$$

By adding these equations and rearranging terms, the intersection point and thus the optimal solution is:

$$\begin{aligned} & kT_{clk} + \Delta - \Delta t_e^o + T_{clk} + t_d + t_{min} - t_{hold} - \Delta t_e^n \\ & = \Delta t_l^n + t_d + t_{max} + t_{setup} + kT_{clk} + \Delta + \Delta t_l^o \\ T_{clk} & = t_{max} - t_{min} + t_{setup} + t_{hold} + \Delta t_l^o + \Delta t_e^o + \Delta t_l^n + \Delta t_e^n \end{aligned}$$

By combining Case 1 and Case 2, the theorem is proved. □

Therefore, it is obvious that the minimum clock period is achievable for any value of k since T_{clk}^* does not depend on k. Note that the clock skew (Δ) needed to achieve this minimum clock period does depend on k as shown in the following theorem.

Theorem 2.3 *The required intentional clock skew (Δ^*) at the optimal clock period (T_{clk}^*) for the single stage system is*

$$(1-k)t_{max}+kt_{min}+(1-k)t_{setup}-kt_{hold}-k\Delta t_l^o+(1-k)\Delta t_e^o+(1-k)\Delta t_l^n-k\Delta t_e^n+t_d$$

for $T_{clk}^* \neq t_{internal}$ or the interval

$$(t_{max}-kt_{internal}+t_d+t_{setup}+\Delta t_l^n+\Delta t_e^o,\ t_{min}+(1-k)t_{internal}+t_d-t_{hold}-\Delta t_l^o-\Delta t_e^n)$$

for $T_{clk}^* = t_{internal}$.

Proof: Note that from the discussion in the proof of Theorem 2.2, it is evident that if $T_{clk}^* = t_{internal}$, the internal node constraint is active. Thus, for the proof of this theorem, there are again two cases depending on where the internal node constraint is active.

Case 1: Internal Node Constraint Not Active ($T_{clk}^* \neq t_{internal}$)
From Theorem 2.2, the minimum clock period is given by:

$$T_{clk}^* = t_{max} - t_{min} + t_{setup} + t_{hold} + \Delta t_l^o + \Delta t_e^o + \Delta t_l^n + \Delta t_e^n. \qquad (22)$$

Also, this minimum clock period must satisfy the constraint 1. Thus by combining equation 22 and the boundary of constraint 1, the required intentional clock skew (Δ^*) must be:

$$\Delta^* = (1-k)t_{max} + kt_{min} + (1-k)t_{setup} - kt_{hold} - k\Delta t_l^o$$
$$+(1-k)\Delta t_e^o + (1-k)\Delta t_l^n - k\Delta t_e^n + t_d$$

Case 2: Internal Node Constraint Active ($T_{clk}^* = t_{internal}$)
When the internal node is active, as shown in Figure 2.10, the optimal intentional clock skew can be found by substituting the optimal clock period, $T_{clk}^* = t_{internal}$, into the boundary equations of constraints 1 and 2. This is will imply the range as stated in the theorem. □

2.5 THE PARAMETER k

2.5.1 Degree of Wave Pipelining

From the constraint space diagram of Figure 2.9, it is evident that the same clock period can be achieved with different values of k. It is also evident that the feasible region is different for different values of k. Therefore it is important to investigate the physical meaning of k.

Along the data path, data that is launched by the input register at time Δ_i is captured by the output register at time $kT_{clk} + \Delta_o$ and new data is launched at time $T_{clk}+\Delta_i$. This implies the following definition for the degree of wave pipelining along the data path (W_d) relative to local clock edges.

$$W_d = \frac{kT_{clk} + \Delta}{T_{clk}}$$

This is the number of clock cycles that elapse between latching the data into the combinational logic and latching the data at the output. Note that, in conventional design where $k = 1$ and Δ=0, $W_d = 1$ as expected.

Similarly, along the clock path, the input clock edge that arrives at the input clock at time Δ_i will arrive at the output register at time Δ_o. Thus, the following definition for the degree of wave pipelining along the clock path (W_c) is made.

$$W_c = \frac{\Delta}{T_{clk}}$$

This is the number of clock cycles elapsed while the clock edge is begin delayed to generate the local clocks at the input and output of the combinational logic block.

Using the definitions for W_d and W_c, k is the global clock latency.

$$k = W_d - W_c$$

This statement shows that k is the difference in the number of waves present in the data chain and the number of waves present in the clock chain. In other words, k is the number of global clock cycles elapsed between launching of data at the input register and capturing the data at the output register.

As an illustration, consider the single stage system of Figure 2.11 and its associated constraint space diagram assuming setup and hold times along with unintentional clock skew are zero. The dashed vertical line indicates an intended clock period of T_{clk}=4. It is evident that, to achieve this clock period, the output register clock skew could be set to Δ=13, Δ=9, Δ=5, or Δ=1 as indicated on the constraint space diagram. Each of these clock skews places operation in a different k region $k = 0$, $k = 1$, $k = 2$, $k = 3$ respectively. These operating points are depicted in Figures 2.12, 2.13, 2.14, and 2.15.

Note that, in these space-time logic diagrams, an additional axis has been added for the global clock. Local clock edges are depicted as solid vertical lines at the input and output and clock delay through clock delay units are depicted as dashed lines between global clock edges and local clock edges. These figures graphically show the relative skew of both clock and data. Note also that, while these figures assume ideal clock delay units indicated by single dashed lines, variation in clock delay units could be indicated by cones bounded by maximum and minimum delays as in the data path.

In Figure 2.12 where $k = 0$, data is captured by a delayed version of the same global clock edge that launched the data indicating a global clock latency of zero. In Figure 2.13 where $k = 1$, data is captured by a delayed version of the global clock edge of the next cycle after the launching edge indicating a global clock latency of one. Note that if k is forced to be one and Δ is forced to be zero, as in the conventional case, the clock period T_{clk} is forced to be greater than t_{max}. In Figure 2.14 where $k = 2$, it is evident that data launched by a given global clock edge will be captured by a delayed version of the global clock edge two cycles later. Figure 2.15 is similar for $k = 3$.

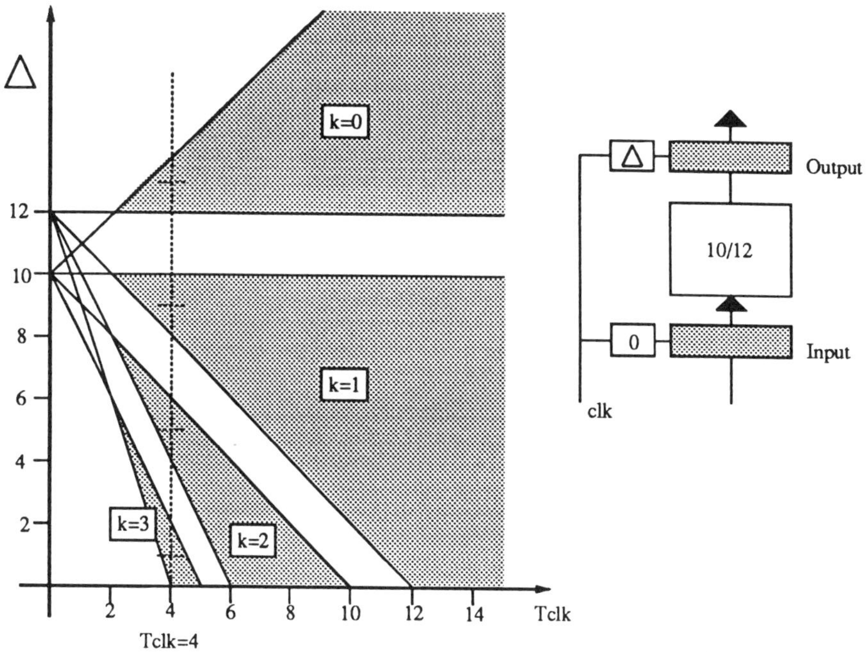

Figure 2.11 Single Stage Example

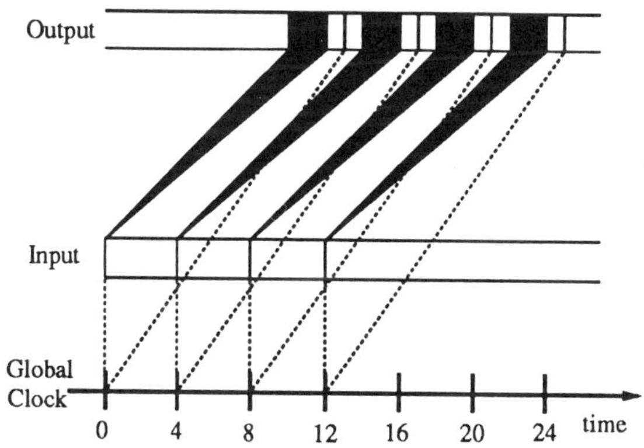

Figure 2.12 Single Stage Example (k=0, Δ= 13)

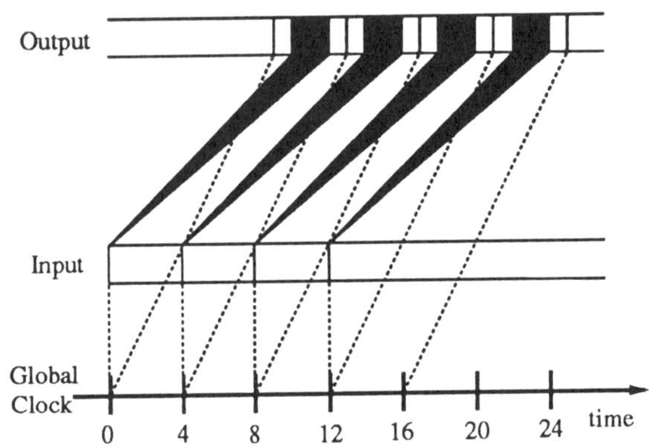

Figure 2.13 Single Stage Example (k=1, Δ= 9)

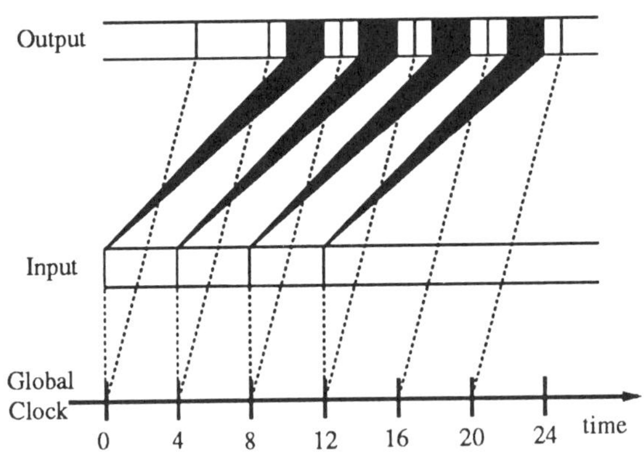

Figure 2.14 Single Stage Example (k=2, Δ= 5)

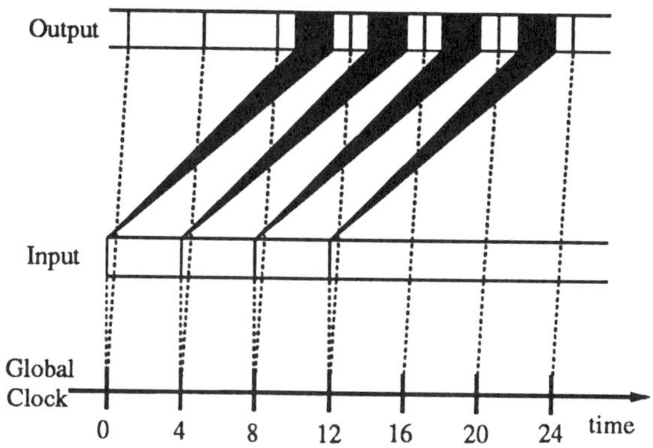

Figure 2.15 Single Stage Example (k=3, Δ= 1)

2.5.2 Frequency Range

From the above example, the question naturally arises as to whether there is any perceivable difference in operation in different k modes. Since k is obviously not equivalent to speedup, different combinations of k and Δ for the single stage system can produce the identical speedup and thus the identical operation from the system point of view.

While this is true for the single stage system at a particular clock period (T_{clk}), there are differences in terms of the range of frequencies over which proper operation can be maintained. For example, consider the constraint space diagram in Figure 2.9. It is obvious that higher k regions have a smaller range of frequencies over which proper clocking is maintained.

Mathematically this can be expressed by considering the register constraints (1) and (2) which are rewritten below in (23) and (24) setting t_{setup}, t_{hold}, and unintentional skew to zero.

$$kT_{clk} + \Delta - \Delta t_e^o \geq \Delta t_l^n + t_d + t_{max} + t_{setup} \tag{23}$$

$$T_{clk} + t_d + t_{min} - t_{hold} - \Delta t_e^n \geq kT_{clk} + \Delta + \Delta t_l^o \tag{24}$$

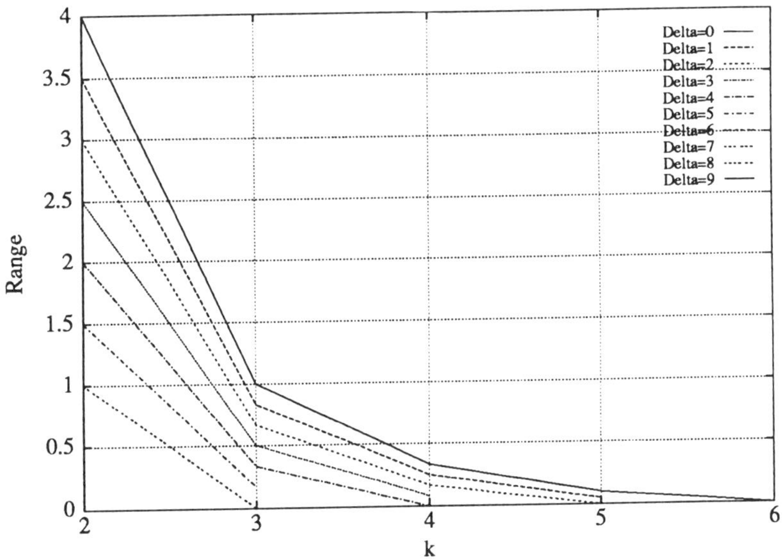

Figure 2.16 T_{clk} Range vs. k (t_{max}=12, t_{min}=10)

By combining these constraints, the bounds on T_{clk} can be determined as:

$$\frac{t_{max} - \Delta}{k} \leq T_{clk} \leq \frac{t_{min} - \Delta}{k - 1} \tag{25}$$

Thus the range of T_{clk} is:

$$\text{range} = \frac{t_{min} - \Delta}{k - 1} - \frac{t_{max} - \Delta}{k} \tag{26}$$

Figures 2.16 and 2.17 graphically depict the relationship between the T_{clk} range and both Δ and k for fixed t_{max}=12 and t_{min}=10.

Further differences in operation for different values of k will become more evident in Chapter 3 when multiple stage pipelines are investigated and in the discussion of the effect of variation on delay in Chapter 6.

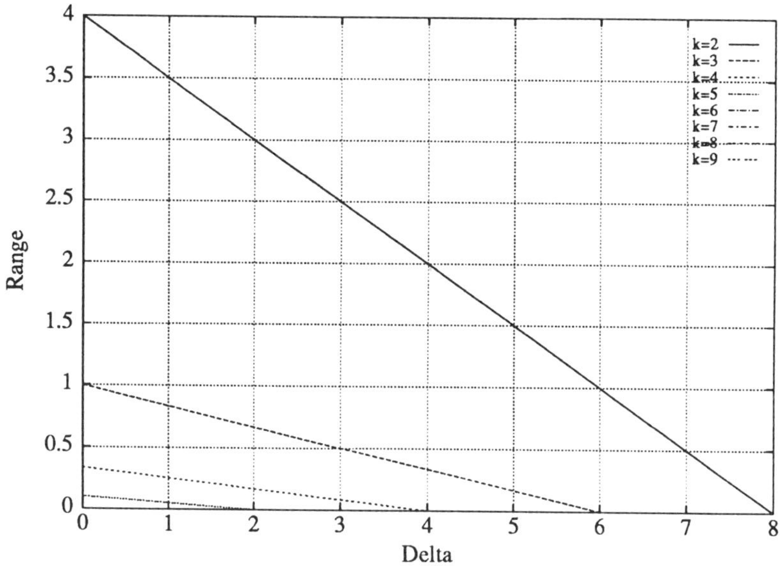

Figure 2.17 T_{clk} Range vs. Δ (t_{max}=12, t_{min}=10)

2.6 SPECIAL CASES

There are several special cases of the general theoretical framework that are interesting from a design point of view. In this section, the effect of each of these special case assumptions on the operation of the wave pipelined system is analyzed. For simplicity in presentation, a single stage system is assumed and unintentional skew, register delay, setup, and hold times are set to zero for this section. Note that these results can easily be extended to the multistage systems and systems with nonzero unintentional skew, register delay, setup, and hold times.

2.6.1 $\Delta \leq T_{clk}$

The addition of this constraint forces the intentional skew to, in effect, only be used for fine tuning the clock period. That is, in order to operate at the smallest clock period, the designer is forced to operate in the highest k region so that, as much as possible, most of the input to output delay is derived from the cyclic nature of the clock with the remainder supplied by the physical delay.

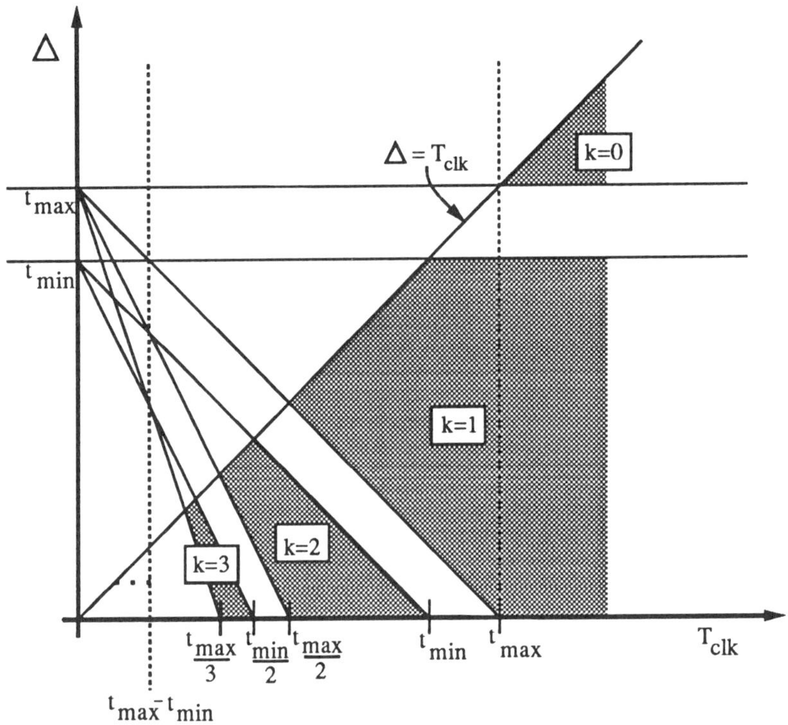

Figure 2.18 Timing Constraint Space Diagram ($\Delta \leq T_{clk}$)

From the following theorem and the constraint space plot shown in Figure 2.18, it is evident that the optimal clock period (T^*_{clk}) is now dependent on k under certain conditions.

Theorem 2.4 *The minimum clock period (T^*_{clk}) of the single stage system with $\Delta \leq T_{clk}$ is*

$$\max(t_{internal}, \frac{t_{max}}{k+1}, t_{max} - t_{min}). \tag{27}$$

Proof: For this special case, there are four constraints concerning the allowable clock period:

$$kT_{clk} + \Delta \geq t_{max} \tag{28}$$

$$t_{min} \geq (k-1)T_{clk} + \Delta \tag{29}$$

$$T_{clk} \geq t_{internal} \tag{30}$$
$$T_{clk} \geq \Delta \tag{31}$$

Constraint (30) alone implies:

$$T_{clk} \geq t_{internal}$$

Constraints (28) and (29) together imply:

$$T_{clk} \geq t_{max} - t_{min}$$

Constraints (28) and (31) together imply:

$$T_{clk} \geq \frac{t_{max}}{k+1}$$

Constraints (29) and (31) do not imply a bound on T_{clk}. Therefore, since there are no other lower bounds on T_{clk}, the minimum clock period (T^*_{clk}) must the maximum of the three lower bounds as stated in the theorem. □

2.6.2 $\Delta \geq 0$

In the general set of constraints, the introduced clock delay Δ is derived from the physical delay units at the input and output registers (Δ=Δ_o-Δ_i). An alternate system is shown in Figure 2.19 in which a single clock delay is introduced in a cascade fashion between the two registers. This type of system implicitly imposes a lower bound on the introduced clock delay (Δ) and thus imposes an upper limit on the clock period for most k regions. The following theorem gives the range of valid T_{clk} for the single stage system with this additional constraint.

Theorem 2.5 *The minimum clock period (T^*_{clk}) of the single stage system with $\Delta \geq 0$ is*

$$\max(t_{internal}, t_{max} - t_{min}) \tag{32}$$

and the maximum clock period is

$$\frac{t_{min}}{k-1}. \tag{33}$$

Proof:
Minimum Clock Period:

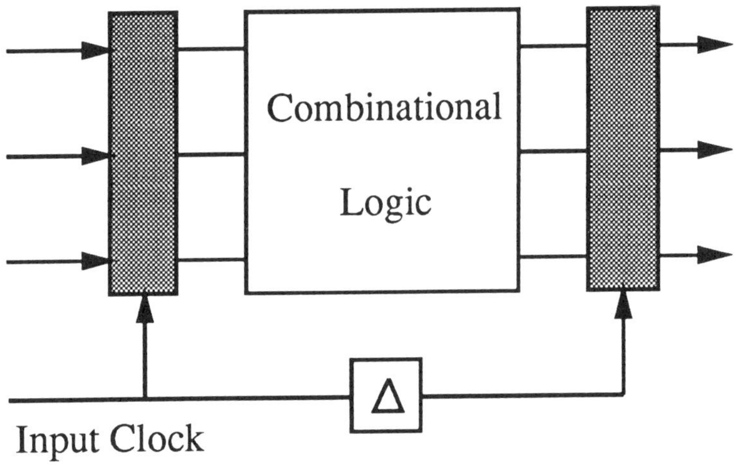

Figure 2.19 Single Stage Block Diagram (Cascaded Clock)

For this special case, there are four constraints concerning the allowable clock period:

$$kT_{clk} + \Delta \geq t_{max} \quad (34)$$
$$t_{min} \geq (k-1)T_{clk} + \Delta \quad (35)$$
$$T_{clk} \geq t_{internal} \quad (36)$$
$$\Delta \geq 0 \quad (37)$$

Constraint (36) alone implies:

$$T_{clk} \geq t_{internal}$$

Constraints (34) and (35) together imply:

$$T_{clk} \geq t_{max} - t_{min}$$

Constraints (34) and (37) do not imply a bound on T_{clk} and constraints (35) and (37) imply an upper bound on T_{clk}. Therefore, since there are no other lower bounds on T_{clk}, the minimum clock period (T^*_{clk}) must the maximum of the two lower bounds as stated in the theorem.

Maximum Clock Period:
From the minimum clock period case, it can be seen that the only combination

of constraints that imply an upper bound on T_{clk} is the combination (35) and (37). These constraints imply:

$$T_{clk} \leq \frac{t_{min}}{k-1}$$

Since there are no other upper bounds on T_{clk}, the maximum clock period is as stated in the theorem. □

2.6.3 $\Delta = 0$

If the clock is routed without intentional skew to both registers, an implicit constraint of $\Delta = 0$ is added. The following theorem gives the range of valid clock period for this case. Note that the valid clocking intervals identified by Lam et. al. in [54] are similar to these special case results.

Theorem 2.6 *The minimum clock period (T^*_{clk}) of the single stage system is*

$$\max(t_{internal}, \frac{t_{max}}{k}). \tag{38}$$

and the maximum clock period is

$$\frac{t_{min}}{k-1}. \tag{39}$$

Proof:
Minimum Clock Period:
Again if the internal node constraint is active, the minimum clock period is $t_{internal}$. Otherwise, the clock period is bounded by constraints (1) and (2). Since constraint (1) is the only lower bound constraints, substituting $\Delta = 0$ in this constraint yields a minimum clock period of $\frac{t_{max}}{k}$.

Maximum Clock Period:
Similar to the minimum clock period case, substituting $\Delta = 0$ in the only upper bound constraint (2) yields a maximum clock period of $\frac{t_{min}}{k-1}$. □

2.6.4 $k = 0, k = 1$

Two interesting special cases are when $k = 0$ and when $k = 1$. In these cases, the range of clock periods is not bounded above even with Δ fixed as can be seen from Figure 2.9. This implies that after Δ is fixed, the system can be run at any clock period from the minimum clock period to infinity. This kind of system is referred to as ***single steppable***. The $k = 1$ case was formulated by Fishburn in [26]. Operation in this mode alleviates the debugging problems that are normally associated with wave pipelined circuits in that the range of frequencies is continuous above the minimal clock period as in conventional circuits. However, a larger Δ is required to obtain the minimal clock period in these modes. As is evident from Figure 2.9, this delay is on the order of the length of the entire combinational logic depth (t_{min} for k=1 and t_{max} for k=0).

If k=1 and Δ=0, the system is a conventional pipeline system and by Theorem 2.6 the clock period is limited by the maximum delay plus setup time and unintentional clock skew as expected.

2.7 CONCLUSIONS

This chapter has presented a rigorous theoretical framework for the analysis of single stage wave pipelined circuits. Timing constraints for the correct clocking have been given and it has been shown that the clock period is bounded below by the difference in the longest (t_{max}) and shortest (t_{min}) delays of the combinational logic block. It has also been shown that distinct regions of operation are available for correct clocking of wave pipelined systems. The size of these regions, in terms of the range of valid clock periods and the range of valid clock skews, is dependent on the global clock latency, k. The minimum clock period is obtained by clock skew optimization formulated as a linear program. In addition to the generalized system, important special cases were examined, and their relative performance limits were analyzed.

In the next chapter, this theoretical framework is extended to the general multistage sequential circuit allowing for the possibility of feedback loops.

3

CLOCK PERIOD CONSTRAINTS: MULTIPLE STAGE SYSTEMS

3.1 INTRODUCTION

In the last chapter the single stage timing constraints of the single stage system were examined in detail. In this chapter, these results are extended to the general case of the multistage system with possible feedback around registers. Most of the theory developed in the last chapter can easily extended to the multistage system. The goal of this chapter is to formalize the constraints of the general case and to present new performance results that are derived from this theoretical formulation.

This chapter is organized into five sections. Section 3.2 defines the system model used for this chapter and extends the notation of the previous chapter to the multistage case. Section 3.3 develops the constraints for correct clocking including the additional constraint involving feedback loops. This section also presents new results involving the additional global limitation placed on the clock period as a result of feedback loops. Section 3.4 then gives a full example of the minimization of the clock period by clock skew optimization for a multistage system with feedback. Here, it is demonstrated how the global constraints can limit the clock period to considerably less than the t_{max}-t_{min} limit. Lastly, Section 3.5 gives conclusions and summaries.

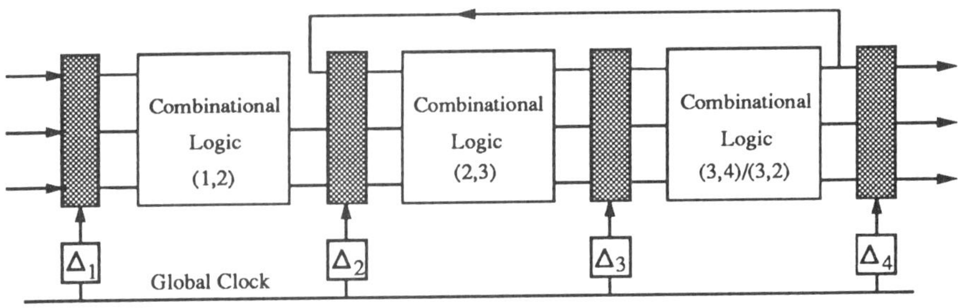

Figure 3.1 Multi-Stage Block Diagram

3.2 SYSTEM MODEL

3.2.1 Circuit Model

A representative block diagram of the multi-stage system to be considered is shown in Figure 3.1. This figure shows a three stage system surrounded by edge-triggered registers with feedback around two of the stages. In general, a system with n edge-triggered registers separating disjoint blocks of combinational logic is assumed. Each combinational logic block is labeled with an ordered pair (i, j) where i represents its input register and j represents its output register. As in Chapter 2, it is assumed that all input bits of each combinational logic block are synchronized by a single local clock. These local clocks are derived from a global clock through delay units (represented as Δ_1 through Δ_4 in Figure 3.1).

Note that the common finite state machine model can be equivalently represented as a general n stage system, and therefore results derived here will be equally applicable to those systems. Also, the restriction of synchronized bits in each register can be easily relaxed to include a system with independent delay units for each register bit.

The combinational logic blocks are described as networks of circuit elements as for the single stage system using the defined sets $\mathcal{N}_{i,j}$, $\mathcal{G}_{i,j}$, and $\mathcal{C}_{i,j}$ subscripted to indicate particular combinational logic blocks. An additional set $\mathcal{R}$ is defined to represent the set of connections between registers (equivalently, the set of combinational logic block).

In summary, the following notation will be used to represent the multi-stage system:

$\mathcal{R}$	set of all register pairs surrounding combinational logic blocks
$\mathcal{N}_{i,j}$	set of all circuit nodes in logic block between registers i and j
$\mathcal{G}_{i,j}$	set of all gate output nodes in logic block between registers i and j
$\mathcal{C}_{i,j}$	set of all circuit connections in logic block between registers i and j

3.2.2 Timing Model

The circuit, clock, and register timing parameters are similarly defined as in Chapter 2. Parameters are superscripted to indicate logic block association. Note that for the multi-stage system, each register has a separate delayed local clock. These delays are represented by Δ_i for each register i.

The notation used for all of the timing parameters for the multi-stage system is listed below.

$t_{max}^{(i,j)}$	maximum delay from register i to register j
$t_{min}^{(i,j)}$	minimum delay from register i to register j
$t_{max}^{(i,j)}(x)$	maximum delay from register i to internal node x in logic block (i,j)
$t_{min}^{(i,j)}(x)$	minimum delay from register i to internal node x in logic block (i,j)
$t_{stable}^{(i,j)}(x)$	minimum stable time for inputs to driving gate(s) of internal node x in logic block (i,j)
T_{clk}	clock period
Δ_i	intentional delay added to clock at register i
Δt_e^i	earliest possible clock skew at register i
Δt_l^i	latest possible clock skew at register i
$t_d(i)$	delay time of register i
$t_{setup}(i)$	set up time of register i
$t_{hold}(i)$	hold time of register i.

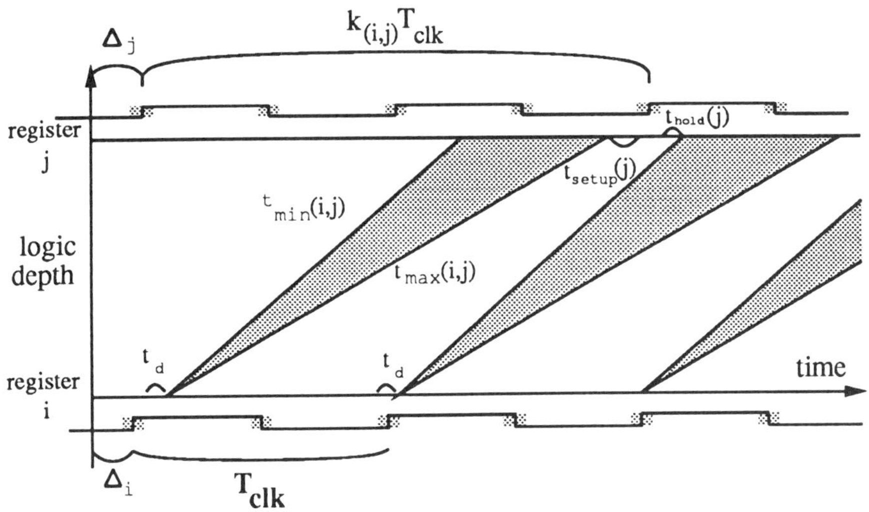

Figure 3.2 Constraints at Registers

3.3 CONSTRAINTS FOR CORRECT CLOCKING

3.3.1 Register Constraints

As with the single stage system, timing requirements must be met at all the registers of the system to insure proper operation. That is, data at the output of a combinational logic block must be captured after the latest present data arrives and before the earliest subsequent data arrives as shown in Figure 3.2. These timing requirements must be met for all pairs of registers surrounding combinational logic blocks ($\forall(i,j) \in \mathcal{R}$). Inequalities (1) and (2) formalize these constraints. Similar to the parameter k of the single stage formulation, $k_{(i,j)}$ is the global clock latency for combinational logic block (i,j).

$$k_{(i,j)}T_{clk} + \Delta_j - \Delta t_e^j \geq \Delta_i + \Delta t_l^i + t_d + t_{max}^{(i,j)} + t_{setup}(j), \quad \forall(i,j) \in \mathcal{R} \quad (1)$$

$$\Delta_i - \Delta t_e^i + t_d + T_{clk} + t_{min}^{(i,j)} - t_{hold}(j) \geq k_{(i,j)} T_{clk} + \Delta_j + \Delta t_l^j, \quad \forall (i,j) \in \mathcal{R} \quad (2)$$

By combining (1) and (2), the following statement must be true.

$$T_{clk} \geq t_{max}^{(i,j)} - t_{min}^{(i,j)} + t_{setup}(j) + t_{hold}(j) + \Delta t_e^i + \Delta t_e^j + \Delta t_l^i + \Delta t_l^j, \quad \forall (i,j) \in \mathcal{R} \quad (3)$$

$$T_{clk} \geq \max_{\forall (i,j) \in \mathcal{R}} (t_{max}^{(i,j)} - t_{min}^{(i,j)} + t_{setup}(j) + t_{hold}(j) + \Delta t_e^i + \Delta t_e^j + \Delta t_l^i + \Delta t_l^j) \quad (4)$$

This implies that the minimum clock period (T_{clk}) is limited by the worst case combinational block in terms of variation in delay, register parameters, and unintentional skew. This reaffirms the importance of delay balancing in *all* combinational logic blocks.

3.3.2 Internal Node Constraints

In addition to the register constraints, timing constraints must be met at all internal nodes of the system.

$$T_{clk} + \min_{(i,j) \in C} (t_{min}^{(u,v)}(i)) - \Delta t_e^u \geq \max_{(i,j) \in C} (t_{max}^{(u,v)}(i)) + \Delta t_l^u + t_{stable}^{(u,v)}(j), \quad \forall j \in \mathcal{G}_{u,v}, \forall (u,v) \in \mathcal{R} \quad (5)$$

Similar to the register constraints, the clock period (T_{clk}) will ultimately be limited by the worst case circuit element in any combinational block.

3.3.3 Feedback Constraints

In order to maintain proper system operation, data initiated on different clock cycles at the system inputs must never collide. In a system without feedback, this is obviously true if valid clocking constraints are satisfied in each individual stage. However, in systems with feedback, an additional constraint must be met at the feedback points. That is, data arriving from the feedback path must arrive with the proper data from the forward path. If for a given feedback loop l in the original system, there are n_l registers in the loop, the number of physical pipeline stages around the loop is n_l. The goal after optimization is to increase the number of *effective* pipeline stages by some factor $\overline{k}$ so that the number of effective pipeline stages is $\overline{k}n_l$. $\overline{k}$, then, is the overall system speedup factor and must be constant for all loops to maintain proper operation. The following theorem formalizes these conditions under which proper operation is insured. Note that this theorem is similar to Theorem 20 of [22] arrived at by a different formulation.

Theorem 3.1 *For a general system with valid clocking in each individual stage, proper system operation is maintained if*

$$\frac{\sum_{\forall (i,j) \in l} k_{(i,j)}}{n_l} = \overline{k} \, \forall l \in \mathcal{S}$$

where $\overline{k}$ is the overall system speedup factor, $\mathcal{S}$ is the set of all loops in the system and n_l is the number of registers in loop l.

Proof: Assume that the goal is to speed up the entire system by a factor $\overline{k}$. Initially, before speedup, the delay around a particular feedback loop (l) is n_l clock cycles where n_l is the number of registers in the feedback loop. In order for proper operation from a global system point of view after speedup, the number of clock cycles around any feedback loop (l) must be $\overline{k}n_l$ so that data coherency is maintained.

The total delay around a feedback loop is the sum of the difference between the input and output clocks for all stages in the feedback loop. For stage (i, j) this delay is $k_{(i,j)}T_{clk} + \Delta_j - \Delta_i$. Therefore, around any particular feedback cycle (l) (as shown in Figure 3.3) the delay (d_n) is:

$$\begin{aligned} d_n &= \sum_{i=2}^{n}(k_{(i-1,i)}T_{clk} + \Delta_i - \Delta_{i-1}) + k_{(n,1)}T_{clk} + \Delta_1 - \Delta_n && (6) \\ &= k_{(1,2)}T_{clk} + \Delta_2 - \Delta_1 + k_{(2,3)}T_{clk} + \Delta_3 - \Delta_2 + \cdots + \\ &\quad k_{(n-1,n)}T_{clk} + \Delta_n - \Delta_{n-1} + k_{(n,1)}T_{clk} + \Delta_1 - \Delta_n && (7) \\ &= \sum_{\forall (i,j) \in l} k_{(i,j)}T_{clk}. && (8) \end{aligned}$$

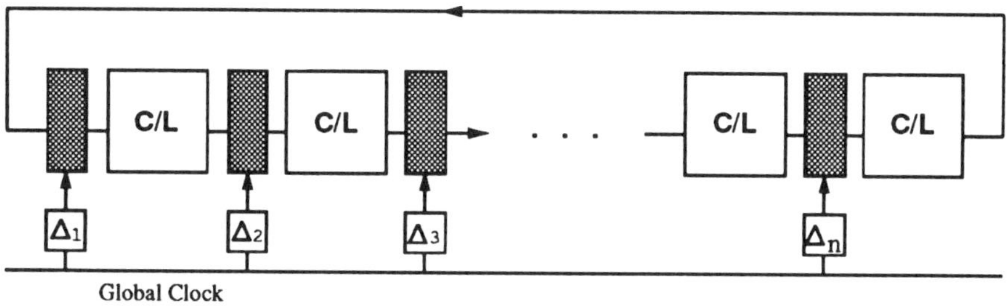

Figure 3.3 n-Stage Feedback

Thus the total number of clock cycles it takes to traverse the feedback cycle is:

$$\frac{d_n}{T_{clk}} = \frac{\sum_{\forall(i,j)\in l} k_{(i,j)} T_{clk}}{T_{clk}} \tag{9}$$

$$= \sum_{\forall(i,j)\in l} k_{(i,j)}. \tag{10}$$

Thus the speedup of the new system over the original system is

$$\frac{\sum_{\forall(i,j)\in l} k_{(i,j)}}{n_l} = \overline{k}$$

or the average k value around the loop. The value must be constant for all loops to maintain proper operation. □

This theorem implies that the average k value ($\overline{k}$) of every loop must be the same to insure the same speedup in all loops. Notice however if the same k is used in all stages of the system, this condition is automatically satisfied and the overall speedup is just k. In this case satisfaction of the register constraints alone will insure proper system operation. Section 3.3.5 will discuss the uniform k system further.

3.3.4 Minimizing the Clock Period

As in Chapter 2, given the constraints identified in the previous section, a minimization problem to minimize the clock period (T_{clk}) can now be formulated. Given variable T_{clk} and Δ_i for all registers i and all other parameters fixed, the following linear program is formulated. As stated in Section 3.3.3, values for $k_{(i,j)}$ must be chosen to satisfy Theorem 3.1.

Min: T_{clk}

Subject to:

$$k_{(i,j)}T_{clk} + \Delta_j - \Delta t_e^j \geq \Delta_i + \Delta t_l^i + t_d + t_{max}^{(i,j)} + t_{setup}(j),\ \forall(i,j) \in \mathcal{R}$$

$$\Delta_i - \Delta t_e^i + t_d + T_{clk} + t_{min}^{(i,j)} - t_{hold}(j) \geq k_{(i,j)}T_{clk} + \Delta_j + \Delta t_l^j,\ \forall(i,j) \in \mathcal{R}$$

$$T_{clk} + \min_{(i,j)\in C}(t_{min}^{(s,t)}(i)) - \Delta t_e^s \geq \max_{(i,j)\in C}(t_{max}^{(s,t)}(i)) + \Delta t_l^s + t_{stable}^{(s,t)}(j),$$

$$\forall j \in \mathcal{G}_{s,t}, \forall(s,t) \in \mathcal{R}$$

3.3.5 Interpretation of k

In Chapter 2, it was shown that speedup was not directly related to k for the single stage system. For the multistage system *without* feedback, this is also the case since the difference in clock delays to the output and input registers (Δ_j-Δ_i) is an independent variable for each combinational logic block and thus can be determined independently.

However, the addition of feedback imposes an additional constraint on the relationship of the difference in delays. This fact manifests itself in additional implied constraints on the clock period (T_{clk}) in systems with feedback. This additional constraint is demonstrated in the following theorem. For notational convenience, unintentional skew, setup, and hold times are set to zero.

Theorem 3.2 *Let $\mathcal{S}$ be defined as the set of all loops in the system, then for multistage systems with feedback*

$$\frac{\sum_{\forall(i,j)\in l} t_{min}^{(i,j)}}{\sum_{\forall(i,j)\in l}(k_{(i,j)} - 1)} \geq T_{clk} \geq \frac{\sum_{\forall(i,j)\in l} t_{max}^{(i,j)}}{\sum_{\forall(i,j)\in l} k_{(i,j)}} \forall l \in \mathcal{S}. \tag{11}$$

Proof: In order to prove that the above constraints are satisfied for all loops in $\mathcal{S}$, it must be proven that the constraints are satisfied for any general loop in the system. To this end, consider the general n stage loop shown in Figure 3.3. Rewriting lower bound constraint (1) of Chapter 2 for each stage in the feedback loop yields the following set of constraints.

$$k_{(1,2)}T_{clk} + \Delta_2 - \Delta_1 \geq t_{max}^{(1,2)} \tag{12}$$

$$k_{(2,3)}T_{clk} + \Delta_3 - \Delta_2 \geq t_{max}^{(2,3)} \tag{13}$$

$$k_{(3,4)}T_{clk} + \Delta_4 - \Delta_3 \geq t_{max}^{(3,4)} \tag{14}$$

$$\vdots$$

$$k_{(j-1,j)}T_{clk} + \Delta_j - \Delta_{j-1} \geq t_{max}^{(j-1,j)} \tag{15}$$

$$\vdots$$

$$k_{(n-1,n)}T_{clk} + \Delta_n - \Delta_{n-1} \geq t_{max}^{(n-1,n)} \tag{16}$$

$$k_{(n,1)}T_{clk} + \Delta_1 - \Delta_n \geq t_{max}^{(n,1)} \tag{17}$$

Combining constraints (12) and (13) yields:

$$\Delta_3 \geq t_{max}^{(1,2)} + t_{max}^{(2,3)} - k_{(1,2)}T_{clk} - k_{(2,3)}T_{clk} + \Delta_1 \tag{18}$$

Now, combining constraints (18) and (14) and so forth for all other constraints:

$$\begin{aligned}\Delta_4 \geq\ & t_{max}^{(1,2)} + t_{max}^{(2,3)} + t_{max}^{(3,4)} - k_{(1,2)}T_{clk} - k_{(2,3)}T_{clk} \\ & -k_{(3,4)}T_{clk} + \Delta_1\end{aligned} \tag{19}$$

$$\vdots$$

$$\Delta_j \geq \sum_{i=2}^{j} t_{max}^{(i-1,i)} - T_{clk} \sum_{i=2}^{j} k_{(i-1,i)} + \Delta_1 \tag{20}$$

$$\vdots$$

$$\Delta_n \geq \sum_{i=2}^{n} t_{max}^{(i-1,i)} - T_{clk} \sum_{i=2}^{n} k_{(i-1,i)} + \Delta_1 \tag{21}$$

Finally combining constraints (21) and (17) and rearranging terms, constraint (22) is implied.

$$T_{clk} \sum_{\forall (i,j) \in l} k_{(i,j)} \geq \sum_{\forall (i,j) \in l} t_{max}^{(i,j)} \tag{22}$$

Similarly by combining the upper bound constraints (2) of Chapter 2 for each stage in the feedback loop yields the following constraint.

$$T_{clk} \sum_{\forall (i,j) \in l} (k_{(i,j)} - 1) \leq \sum_{\forall (i,j) \in l} t_{min}^{(i,j)} \tag{23}$$

By combining constraints (22) and (23), the constraint given in Theorem 3.2 is implied (for $\sum(k_{(i,j)} - 1) > 0$).

$$\frac{\sum_{\forall(i,j)\in l} t_{min}^{(i,j)}}{\sum_{\forall(i,j)\in l}(k_{(i,j)} - 1)} \geq T_{clk} \geq \frac{\sum_{\forall(i,j)\in l} t_{max}^{(i,j)}}{\sum_{\forall(i,j)\in l} k_{(i,j)}}. \tag{24}$$

Since the above constraint must be satisfied for any general cycle in the system, the theorem is proved. □

Any multistage system (as shown in Figure 3.4) can be viewed globally as a single stage system with timing parameters $\mathbf{TMAX} = \sum_{\forall(i,j)\in\mathcal{R}} t_{max}^{(i,j)}$, $\mathbf{TMIN} = \sum_{\forall(i,j)\in\mathcal{R}} t_{min}^{(i,j)}$, and $\mathbf{K} = \sum_{\forall(i,j)\in\mathcal{R}} k_{(i,j)}$. Constraints analogous to constraints (2) and (1) of Chapter 2 can be written for the global system (see Figure 3.5) where n is the number of stages in the system.

$$\mathbf{K}T_{clk} + \Delta_o \geq \Delta_i + \mathbf{TMAX} \tag{25}$$

$$\Delta_i + nT_{clk} + \mathbf{TMIN} \geq \mathbf{K}T_{clk} + \Delta_o \tag{26}$$

Combining these constraints, a constraint analogous to constraint (4) of Chapter 2 is derived.

$$T_{clk} \geq \frac{\mathbf{TMAX} - \mathbf{TMIN}}{n} \tag{27}$$

With the addition of feedback, the sum of Δ_i around the loop must be zero to insure correct clocking. Thus the multistage system with feedback is analogous to the single stage system with $\Delta = 0$ as described in Section 2.6.3. Rewriting the constraints of Theorem 3.2, the similarity between these constraints and the constraints of Theorem 2.6 can be seen.

$$\frac{\mathbf{TMIN}}{\mathbf{K} - n_l} \geq T_{clk} \geq \frac{\mathbf{TMAX}}{\mathbf{K}} \forall l \in \mathcal{S} \tag{28}$$

In practical terms this theorem shows that, in contrast to the single stage system (Theorem 2.2), the minimum clock period (T_{clk}) for the multistage system with feedback *is* related to k. Each individual stage (i,j) still limits the clock period to the difference in its maximum and minimum ($t_{max}^{(i,j)}$-$t_{min}^{(i,j)}$) and by its internal node constraints. But globally, since the output clock of one stage is the input clock of the next stage, a feedback loop produces a set of circular dependencies that imply a global constraint on the clock period. This limit is the total delay of the feedback loop divided by the total number of global clock cycles that

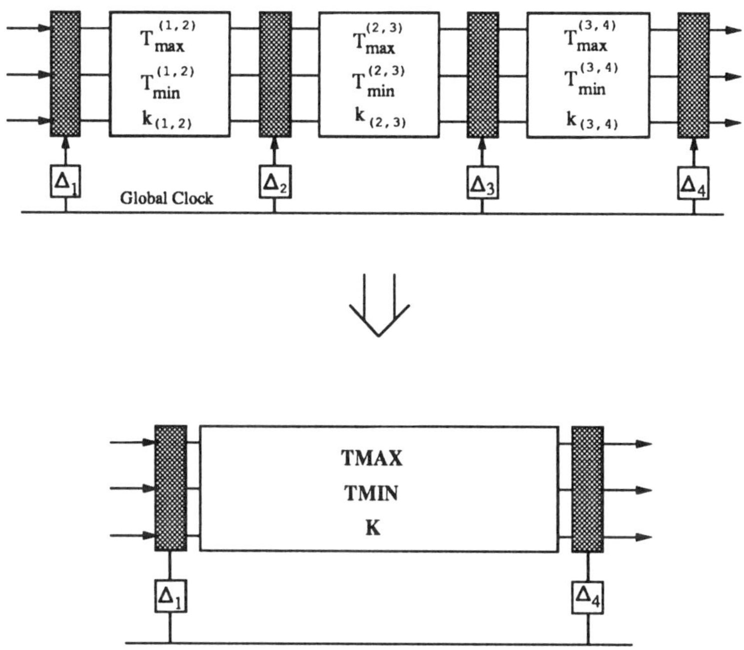

Figure 3.4 Multistage to Single Stage Transformation

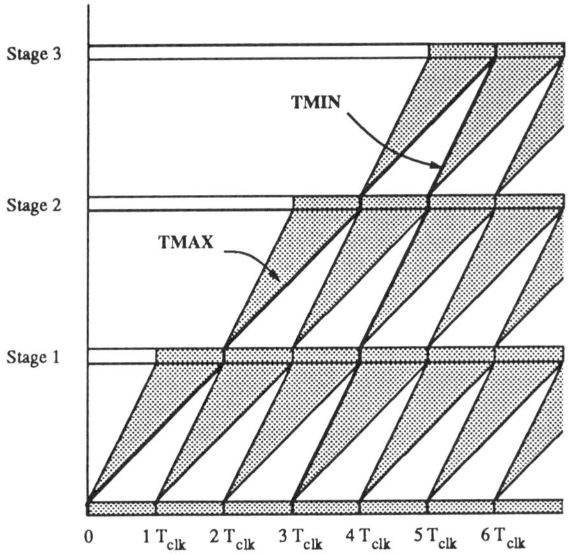

Figure 3.5 Constraints for Multistage System

it takes to get data around the loop. The total delay is just the sum of the maximum delays of each stage in the loop. The total number of global clock periods around the loop is just the sum of the $k_{(i,j)}$ of each stage of the loop since $k_{(i,j)}$ for a particular stage is the number of globals clock cycles to get data through that stage (refer to Section 2.5).

This implies that the only way to increase the speed of a wave pipelined system with feedback in which the global constraint is the limiting constraint is to increase the sum of the $k_{(i,j)}$s around all loops. However, since the sum of $k_{(i,j)}$s also relates to the speedup of the system (Theorem 3.1), changing this sum will change the global operation of the pipeline thus producing a different computation. Whether or not this is allowable depends on overall system level considerations. Note also that increasing the sum provides no advantage in terms of clock period after the maximum rate limits of individual combination logic blocks have been reached.

Also, from Theorem 3.2, this lower bound on the clock period depends on the *sum* of the $k_{(i,j)}$ not the individual values of $k_{(i,j)}$ in each stage. This implies that different combinations of $k_{(i,j)}$ will give the same lower bound. An integer speedup (s) over the non-wave pipelined system can be accomplished by using uniform $k_{(i,j)} = s$ in all stages. This will automatically satisfy Theorem 3.1 to insure correct operation. Note that no other combination of $k_{(i,j)}$s will give a lower bound on the clock period while achieving this speedup. Therefore there is no advantage, in terms of clock period, to use any combination of $k_{(i,j)}$ other than uniform $k_{(i,j)}$ in all stages to yield the given speedup. However, different combinations will require different size delay units (Δ_i) and will yield different frequency ranges analogous to the single stage system described in Sections 2.4 and 2.5.2. If non-integer speedup is desired or if additional limits are placed on delay unit sizes or frequency ranges, non-uniform $k_{(i,j)}$ may be necessary.

3.4 EXAMPLE

As an example of the optimization of a multistage system with feedback, consider the system shown in Figure 3.1. This is a system with three combinational logic blocks surrounded by four registers with feedback from the output of logic block (3,4) to the input of register 2. The delay properties of each of the combinational logic blocks are shown in the following table. Register parameters t_{setup} and t_{hold} along with unintentional skew for all registers is set to zero for this example and, for notational convenience, it is assumed that the internal node requirement is satisfied for all combinational logic blocks. Uniform $k_{(i,j)}$=1 operation is assumed in all stages.

Register Pair	t_{min}	t_{max}
1,2	20	22
2,3	26	27
3,4	2	3

The minimum clock period (T_{clk}) along with the optimal clock skew at each register can be found by the solution of a linear program as outlined in Section 3.3.4. The linear program is show below along with optimal T_{clk} and clock delays.

Min: T_{clk}

Subject to:

$$\begin{aligned} \Delta_2 - \Delta_1 &\leq 20 \\ \Delta_1 - \Delta_2 - T_{clk} &\leq -22 \end{aligned}$$

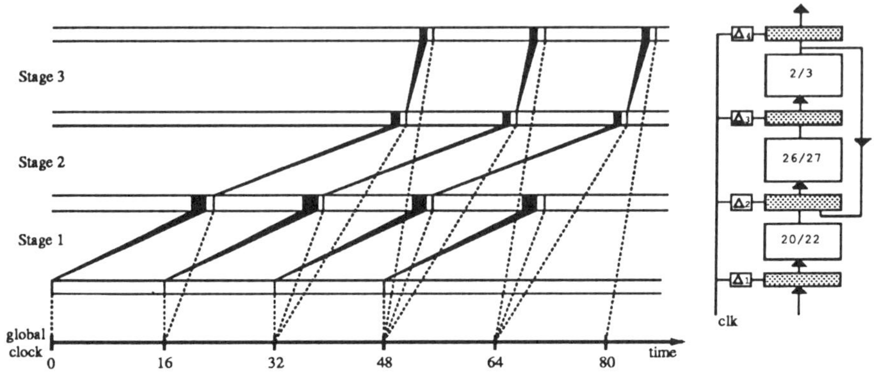

Figure 3.6 Three Stage Example

$$
\begin{aligned}
\Delta_3 - \Delta_2 &\leq 26 \\
\Delta_2 - \Delta_3 - T_{clk} &\leq -27 \\
\Delta_4 - \Delta_3 &\leq 2 \\
\Delta_3 - \Delta_4 - T_{clk} &\leq -3 \\
\Delta_2 - \Delta_3 &\leq 2 \\
\Delta_3 - \Delta_2 - T_{clk} &\leq -3
\end{aligned}
$$

Optimal Solution:

$$
\begin{aligned}
\Delta_1 &= 0 \\
\Delta_2 &= 7 \\
\Delta_3 &= 19 \\
\Delta_4 &= 7 \\
T^*_{clk} &= 15
\end{aligned}
$$

This clock delay solution is shown graphically in Figure 3.6. This timing diagram is shown with T_{clk}= 16 (for clarity the clock period is shown longer than the optimal T^*_{clk}=15).

Since $t^{(i,j)}_{max}$ - $t^{(i,j)}_{min}$ is at most 2 (logic block (1,2)) it is evident that, without feedback, this system could be operated with a clock period of 2. However,

Table 3.1 Linear Program Solution for Different Values of $k_{(i,j)}$

$k_{(i,j)}$	T^*_{clk}	Δ_1	Δ_2	Δ_3	Δ_4
1	15	0	7	19	7
2	7.5	0	7	19	7
3	5	0	7	19	7
4	3.75	0	7	19	7
5	3	0	7	19	7
6	2.5	0	7	19	7
7	2.14	0	7	19	7
8	2	0	6	18	6

with feedback, the period is limited to 15. For comparison, zero skew conventional operation would yield a minimum clock period of at least 27 due to the maximum delay of logic block (2,3), and a system in which the combinational logic between registers 2 and 4 is not pipelined at all would have a minimal clock period of 30.

Table 3.1 gives optimal clock periods and clock skews for the example system with different values of $k_{(i,j)}$. As can be seen, the theoretical limit can be reached with $k_{(i,j)}$=8 in all stages. However, this will obviously place a more stringent requirement on the precision of the delay units and the operating frequency since less variation can be tolerated.

3.5 CONCLUSIONS

This chapter has extended the formulation of Chapter 2 to the multistage system with feedback. For such systems with feedback, it has been shown that feedback loops impose additional implied constraints on the minimal and maximal clock period to insure proper operation at the feedback points. These constraints are dependent on the number of register stages in the feedback loop and the global clock latency. These constraints may limit the operation of a wave pipelined system to significantly less than the t_{max}-t_{min} limit.

While the formulation of this chapter is based on the multistage pipelined model (Figure 3.1), the results can be easily extended to the finite state machine model (Figure 3.7) since the state machine model can be easily transformed into the

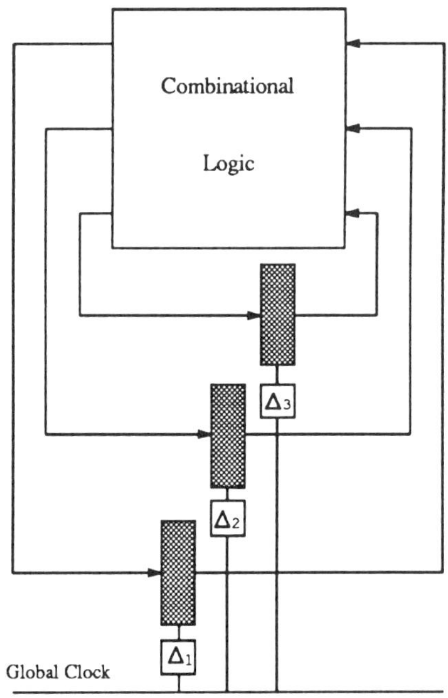

Figure 3.7 Multi-Stage Block Diagram - Finite State Machine

multistage model. However, with global clock latency greater than one ($k > 1$), care must be taken to properly schedule external inputs to the machine so that correct system level operation is insured. For the multistage pipeline model, this relates to the optimal scheduling of conventional multistage pipelines.

The constraints developed in this chapter can be used by designers to make tradeoffs between timing parameters such as the clock period (T_{clk}) and physical parameters such as intentional clock skew (Δ) to meet timing and reliability requirements thus allowing for maximal rate operation in practical multistage systems.

4

EXACT TIMING ANALYSIS

4.1 INTRODUCTION

Chapters 2 and 3 developed a theoretical formulation of the constraints for correct clocking of wave pipelined systems. Throughout the formulation, circuit timing parameters such as the maximum and minimum delays in each combinational block (t_{max} and t_{min}) and the maximum and minimum delays from each register to each internal circuit node ($t_{max}^{(i,j)}(x)$ and $t_{min}^{(i,j)}(x)$) were assumed to be known *a priori*. Based on the necessary constraints for correct clocking, it was shown that the maximum clock frequency directly relates to the difference in maximum and minimum delays through combinational logic blocks. This chapter begins a discussion of methodology and algorithms for accurately determining these critical circuit delays.

Much work has been done in the field of timing verification for conventional systems both in simulating timing behavior and in extracting critical delays. However, maximal performance of wave pipelined systems, as well as high performance systems in general, requires a level of accuracy not generally available in conventional timing analysis tools. This is primarily due to two factors of high performance wave pipelined design. First, timing constraints are generally tight so that inaccurate or overly conservative timing analysis will be of limited use. Second, since the performance of wave pipelined systems depends on the difference in maximum and minimum circuit delays, combinational logic circuits will, in general, be more highly balanced at the gate level. In fact, it is many times feasible to exactly balance path lengths at a graph level such that each path from input to output contains the exact same number of circuit elements. Thus, a main source in overall circuit delay difference arises from data dependent delay variation in individual logic gates.

These concerns point to the need for accurate timing analysis that detects both maximum and minimum delays, reports only delays that can actually be excited by primary input vectors, and is based on accurate circuit element timing models that consider data dependent delay effects. While existing techniques using conventional circuit element delay models can be extended to report minimum as well as maximum delays and eliminate non-sensitizable delays, the delay models used by current timing analysis tools are, in general, not accurate enough for the level of analysis that is needed. As will be shown, sufficiently accurate delay models, in turn, lead to additional concerns in the methodologies used to detect circuit delays and determine sensitization of delays.

As an example of the need for data dependent delay models, consider a circuit composed of two input CMOS NAND gates. As will be shown in Section 4.2.2, the rising delay of a two input CMOS NAND gate can vary by as much as a factor of two based on whether one input is changing or both inputs are changing. Also, both rising and falling delays can vary significantly when only one input is switching depending on which input terminal is changing. This implies that input ordering of circuit elements is important throughout the analysis process and that multiple paths through combinational logic must be examined together in order to determine if simultaneous transitions can occur at individual elements. Circuit element delay models are required that consider delay as a function of rising vs. falling transitions, input terminal ordering, and the difference in arrival times when multiple gate inputs are switching.

In general, for any type of timing analysis to be feasible, a tradeoff must be made trading accuracy for analysis time. Since the purpose of most timing analysis applications is to simply report a bound on the maximum circuit delay, most of the emphasis of current timing analysis work is on reducing runtime for larger circuit analysis. However, for high performance wave pipelined circuit design, accuracy of delay time is also very important. Therefore, this work places more emphasis on accuracy than most previous work and extracts exact delays of combinational circuits.

It is important to note that while the primary target of this timing analysis work is the analysis of wave pipelined circuits, many other types of performance oriented circuit design styles also require very accurate timing analysis. For example, several design styles also require the detection of shortest paths. These include design with single phase latches, asynchronous circuit design, and the design of clock distribution networks [99, 58]. In each of these cases, accurate performance estimation depends on accurate detection of both maximum and minimum circuit delays. Also, even the detection of maximum circuit delay for high performance conventional systems may require determining whether simultaneous or near simultaneous transitions can occur at gate inputs especially when paths are more nearly equalized after delay, area, and power optimization [106] in which maximum delays are shortened to maximize the clock rate and minimum delays are lengthened to minimize power consumption.

The next two chapters of this work present motivation and concerns for the timing analysis of high performance CMOS circuits. Two algorithmic solutions are presented. The first, presented in this chapter, is a relatively straightforward algorithm that is based on fast incremental timing simulation. While this method is feasible for only relatively small circuits due to highly exponential runtime and space requirements, it does demonstrate several important concepts and problems related to accurate timing analysis. In Chapter 5, a more efficient algorithm is presented based on searching more efficient data structures. Limitations and extensions of the overall timing analysis problem are also discussed in that chapter.

In the remainder of this chapter, a brief overview of the field of timing verification is presented. Then, the consideration of minimum delays and more accurate delay models is motivating by presenting an overview of CMOS circuit delay models and implications of these models for critical delay calculations. Next, notations and definitions are presented leading to the development of a basic algorithm for timing analysis.

4.2 MOTIVATION AND JUSTIFICATION

4.2.1 Timing Verification

In general, computer aided techniques for the verification of digital circuit timing fall into two basic categories: simulation and analysis. The first logical technique needed in timing verification is ability to simulate the operation and timing of the circuit. That is, given a particular set of input vectors, determine the response of internal circuits nodes and circuit outputs over time. CAD tools that fall into this category are generally referred to as ***circuit simulators*** or ***timing simulators*** depending on the level of circuit modeling.

Perhaps the most well known members of the circuit simulation category are the SPICE-type simulators. These simulators were derived from University of California at Berkeley research in the mid 1970's. These simulators produce highly accurate results by detailed analysis of currents and voltages at individual circuit nodes over time. The original mathematical techniques and various derivatives of these techniques have been widely used for almost twenty years in many university tools such as the original SPICE2 [72] and SPICE3 [79] circuit simulators as well as commercial and industry circuit simulators such as hSPICE [71] and CAzM [23].

Although these circuit simulators are very accurate, this analysis involves the solution of large systems of ordinary linear differential equations in which the number of equations is related to the number of circuit nodes. Though there have been significant developments in new techniques to reduce runtime and maintain near SPICE-like accuracy using new techniques such as the relaxation techniques [105, 44], asymptotic waveform analysis in AWE [77], and other approximation techniques [89, 94], these techniques are still too slow for the complete verification of large digital systems.

To combat this problem, timing simulators model the circuit as a network of higher level circuit elements with associated timing characteristics. The circuit elements are typically defined either as single transistors modeled as bidirectional resistive switches through which nodal capacitances are charged and discharged or as groups of transistors implementing specific low level logic functions such as AND, OR, NAND, NOR logic gates with associated delays models. Examples of timing simulators include tools such as RSIM [98], IRSIM [88], MOTIS [14], SPECS2 [104], and others [6, 8, 36, 103].

With accurate models for a well characterized technology, timing simulators can produce quite accurate and fast results. However, in order to fully verify the timing of the circuit by determining worst case circuit performance, many simulation runs must be performed using different input vector patterns. In many cases, exhaustive simulation of all possible input vector pairs must be performed. Even with very fast timing simulators based on simplified timing models, it is still only feasible to fully verify very small circuits. This leads to the second category of timing verification tools in which worst case timing is extracted by analyzing the structure of the circuit. These tools are generally referred to as *timing analyzers* or *timing verifiers*. These methods report worst case delay times using an input vector independent approach.

The history of timing verifiers dates back to early PERT techniques [48] developed at IBM in the 1960's. This type of analysis models the circuit as a graph of circuit elements with associated delay values for each node and uses graph techniques to detect longest and shortest paths in the weighted graph. These techniques are very fast and have been used in numerous timing analysis tools since that time. Throughout the 1980's, many tools were developed with increasingly improved circuit timing models and runtimes that remained feasible for reasonable sized circuits. Examples of these types of timing analyzers include TA [32], CRYSTAL [75], TV [38], MOSsim [11], LEADOUT [97], E-TV [47], SUPERCRYSTAL [3], XPSim [4] and others [63, 112, 113, 115].

Though these tools contained increasingly more accurate delays models and can easily find the longest and shortest graph level path lengths, they do not, however, consider whether signals can actually propagate down the selected path. This shortcoming has been labeled the *false path problem*. From the late 1980's through the present, a flurry of activity [9, 7, 76, 21, 67, 16, 19, 91, 34, 20, 54] in the timing analysis area has produced many algorithms and CAD tools to eliminate false paths by detecting which paths can be *sensitized* under various conditions. These methods are compared and contrasted in Section 4.2.3.

In general, there are two basic reasons for performing timing analysis. First is the determination of worst case (maximum and minimum) delays using fixed delay models at design time and reporting vectors that produce these delays. These vectors can then be used in timing simulators or circuit simulators to get precise information about worst case circuits delays. Second is the goal of reporting worst case bounds on the circuit delay for the fabricated circuit over the entire range of process and environmental variation. While actual circuit delays of any fabricated device are guaranteed to fall within the reported bounds, the bounds reported by these timing analyzers are, in general, only tight for the worst case process and environmental conditions and vectors produced by this method are of limited use since they may not be applicable to all process and environmental conditions. While both purposes are important, the primary target for the timing analysis methods proposed in these chapters is to produce exact timing information at design time using fixed delays. However, the proposed techniques can be extended to give worst case information over different process and environmental condition as will be discussed in Section 5.8.2.

4.2.2 Delay Models

In order to perform timing analysis, a circuit is partitioned into a network of base level circuit elements. While some timing analysis tools use individual transistors as these circuit elements[38, 75], most use either simple logic gates (such as NAND, NOR, AND, OR, etc.) or transistor groups. In many cases the design is inherently partitioned into circuit elements such as in design using standard cells or gate arrays. When designs are specified at the transistor level (such as in post-layout verification) transistor groups can be extracted using partitioning techniques such as described in [28]. Using either simple logic gates or transistor groups, delay models for each type of circuit element are then generally derived by intensive circuit level simulation.

In the current critical path literature, most authors assume one of three circuit element delay models to simplify the computation of critical paths. These are ***unit delay*** (each element has a delay of one time unit), ***constant delay*** (a single delay value associated with each element), and ***input order dependent delay*** (a different delay for each input,output pair of the element). In recent work these models have been extended to include rising vs. falling delays [34] and delay ranges to model possible process and environmental variation [91, 95, 54, 70]. However, all of these models assume that the element delay is identical whether only one or multiple element inputs are switching simultaneously.

While this simplification yields near correct results when only long path delays are being considered in standard CMOS logic circuits, it is not adequate when short paths are to be determined. For example, consider the standard static CMOS NAND gate shown in Figure 4.1. It is clear that the longest rising time of the gate will be when one input is held constant at 1 and the other input is switched from 1 to 0. However, the shortest rising time will occur when both inputs are changing from 1 to 0 simultaneously due to the two parallel pullup paths.

It is also evident from the example that gate delay through logic gates can be different for different gate inputs. In the NAND circuit this can be seen by considering the body effect on the serial path transistors [87]. It can be seen that the minimum fall time of the gate will occur when input a is switching from 0 to 1 and input b is held constant at 1. The maximal fall time occurs when both inputs are switching. From the above discussion, it therefore evident that, in order to accurately determine minimum and maximum delays through CMOS logic blocks, relative arrival times of input events and ordering of gate inputs must be considered.

As a first order approximation, RC delay of gate under all input conditions can be used. The resistance of each transistor can be approximated to the first order by assuming linear region operation:

$$R_{tr} = \frac{L/W}{\mu C_{gox}(Vdd - V_T)} \quad (1)$$

This gives a intuitive explanation of the variation in delay possible through the gate.

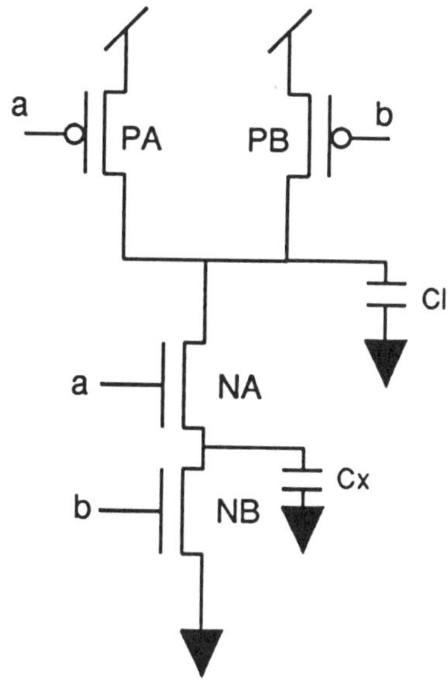

Figure 4.1 Static CMOS NAND Gate

The following equations describe the *falling* delay of the gate based on the RC delay model.

***a* rising, *b* steady**

$$T_{DHL1} = R_{NA}C_l + R_{NB}C_l \tag{2}$$

***a* steady, *b* rising**

$$T_{DHL2} = R_{NA}C_l + R_{NB}(C_x + C_l) \tag{3}$$

***a* rising, *b* rising**

$$T_{DHL3a} = R_{NA}C_l + R_{NB}C_l \qquad (C_x \text{ previously discharged}) \tag{4}$$

$$T_{DHL3b} = R_{NA}C_l + R_{NB}(C_x + C_l) \qquad (C_x \text{ previously charged}) \tag{5}$$

Thus the difference in the maximum and minimum falling delay of the gate is:

$$T_{DHL2} - T_{DHL1} = R_{NB} C_x. \tag{6}$$

The following equations describe the *rising* delay of the gate based on the RC delay model.

a **falling,** b **steady**

$$T_{DLH1} = R_{PA} C_l \tag{7}$$

a **steady,** b **falling**

$$T_{DLH2} = R_{PB}(C_x + C_l) + R_{NA} C_x \tag{8}$$

a **falling,** b **falling**

$$T_{DLH3} = (R_{PA} \| R_{PB}) C_l \tag{9}$$

Thus the difference in the maximum and minimum falling delay of the gate is:

$$T_{DLH2} - T_{DLH3} = (R_{PB} + R_{NA}) C_x + \frac{R_{PB}^2}{R_{PA} + R_{PB}} C_l. \tag{10}$$

These first order equations give an intuitive explanation of gate delay differences, however, they are not sufficiently accurate for detailed delay calculations. For example, in equations T_{DHL1} is shown to be equal to T_{DHL3a} from equations 2 and 4. In reality T_{DHL1} will, in general, be less than T_{DHL3a} since, although transistor NB is modeled the same whether it is switching or stable on, NB begins operation in its linear region when input b is steady while it must switch into its linear region when input b is switching. In addition to the error incurred from the linear region assumption, the simple RC delay models cannot predict the variation in delay based on the relative skew of input transitions. These models have assumed input transitions were simultaneous however, in reality, previous circuits may cause skew between the inputs to the gates and this, in turn, will affect the delay of the gate.

Through detailed circuit level simulation, more exact delay information can be extracted as shown in Figure 4.2. Here, skew between inputs a and b is varied as shown in Figure 4.3 and gate delay is plotted relative to input skew. As can be readily seen, delay varies over a wide range in both rising and falling delay based on whether input transitions are coincident or not. This clearly indicates that, in order to accurately predict maximum and minimum circuit delays,

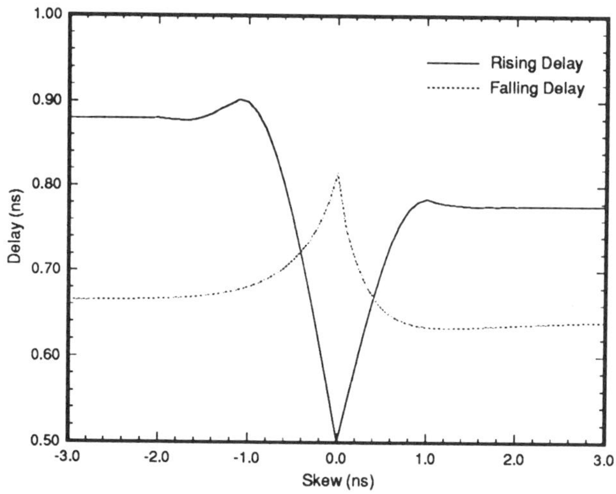

Figure 4.2 Static CMOS Gate Delay Simulations

these effects must be considered. Note that the falling delay curve shown in Figure 4.2 represents that case in which capacitor C_x in Figure 4.1 has been previously discharged. The test circuit configuration and transistor sizes are shown in Figure 4.4.

4.2.3 Sensitization

As previously mentioned, early timing analysis tools [75, 38] calculated critical delays ignoring the functionality of logic gates. That is, any transition on a circuit node is assumed to always be able to pass through the next circuit element. This allows timing analysis by only considering the immediate inputs and outputs of each circuit element. While these assumptions allow for fast timing analysis of logic circuits, this type of timing analysis also potentially reports so called *false paths* that can not be activated by any input pattern. This leads to overly pessimistic approximations of the true circuit delay. If the results of this type of analysis are used to determine which parts of the circuit to optimize, effort may be wasted by unnecessarily optimizing some parts of the circuit. Worse yet, optimization of false paths may inadvertently lengthen the longest true path.

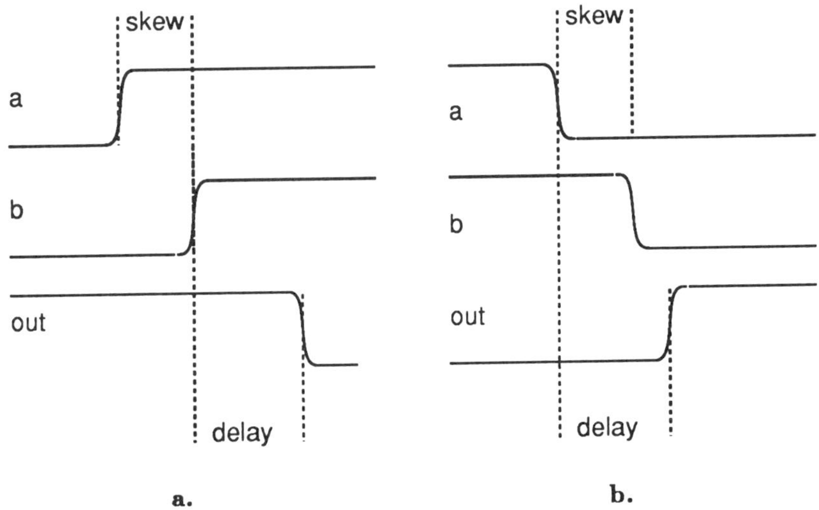

Figure 4.3 Static CMOS Gate Delay Simulations

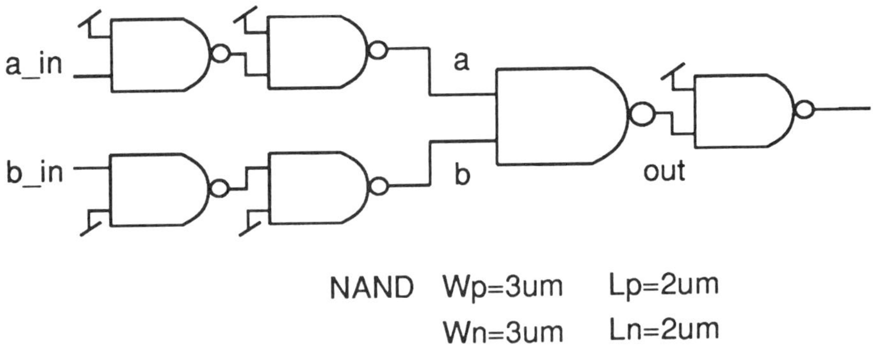

Figure 4.4 Circuit Setup for Skew Simulations

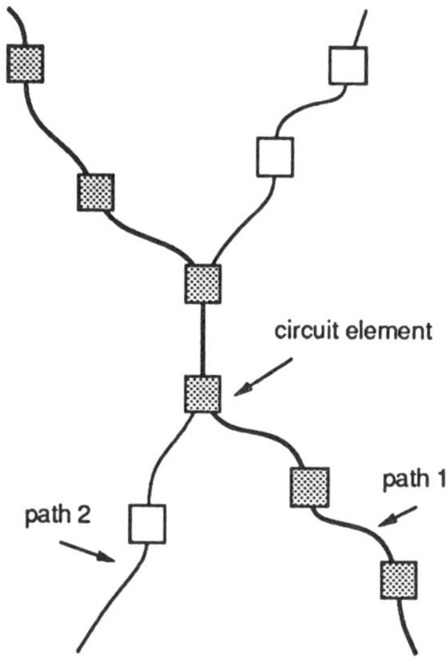

Figure 4.5 Optimization of False Paths

To illustrate the potential lengthening of the longest true path, consider the partial circuit shown in Figure 4.5 with two intersecting paths. Assume path 2 is the longest true path and path 1 is a longer false path. If path 1 is selected for optimization first, transistor sizes will be increased in each gate along the path. This in turn increases the load for driving gates. Thus, the delay of a nonoptimized intersecting path, such as path 2, may increase. This is normally not a problem if the length of the nonoptimized path remains shorter than the original longest path. However, in the case where the optimized path is a *false* path, the optimization process has actually produced a slower circuit.

Because of these concerns, subsequent work [7, 21, 67, 91, 19, 69] has concentrated on elimination of false paths by reporting only paths that can be sensitized under certain sensitizing conditions. In general, the sensitization of a particular path refers to checking if a given transition can be propagated down the path to the output. If so, the path is referred to as a sensitizable path. The goal, then, of these timing analysis packages is to find the longest and shortest ***sensitizable path.***

The underlying principal of path sensitization is that a signal will only pass through a gate if the side inputs of that gate are set to appropriate logic values. That is, the gate is ***sensitized.*** For example, an AND or NAND gate will pass

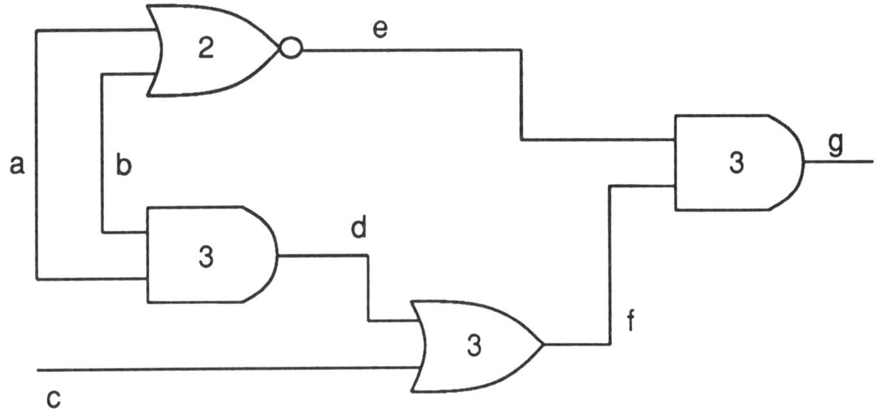

Figure 4.6 Sensitization Example

a switching signal to its output only if all other inputs (the side inputs) are set to logic level 1. Similar, an OR gate will only pass a signal when side inputs are set to 0.

In path sensitization research, there are two basic types of sensitization. First is ***static sensitization*** [7, 43]. In algorithms using this approach, the sensitizing level is set on each of the side inputs of all gates in the path, and it is checked whether or not these conditions can be produced by a single input vector. If so, the path is declared to be statically sensitizable. These methods typically use techniques similar to those used in testability analysis such as the D-algorithm [81] to propagate local constraints at various gates back toward the primary inputs.

To illustrate these sensitization concepts, consider the example circuit shown in Figure 4.6 taken from [9]. The gates models for the AND, OR, and NOR gates using first order delay modeling are shown in Table 4.1. The delay values shown at each gate in Figure 4.6 are the worst case latest delay of each gate used for traditional longest path timing analysis. It is assumed that primary circuit input can only change values at time 0.

The longest physical path of the example circuit is $\{a, d, f, g\}$ with a delay of 9. However, in order for the signal to pass from a to d, b must be set to 1. Also, in order or the signal to pass from f to g, e must be set to 1 implying that b must be set to 0 which conflicts with the previous setting of b. Therefore, it is clear that this path is not statically sensitizable. The static sensitization

Table 4.1 Sensitization Example Gate Models

Input Condition	NOR Delay	AND Delay	OR Delay
2 inputs rising	1	3	2
1 input rising	2	3	3
2 inputs falling	2	2	3
1 input falling	2	3	3

timing analyzer would thus report the delay of $\{c, f, g\}$ (6) as the longest "true" delay since this is the longest path that can be statically sensitized. The input condition that sensitizes this path is $a = b = 0$.

However, there are circuits in which the delay of the longest statically sensitizable path is not the delay of the latest event that can occur at the outputs. This condition can occur on a path that is not statically sensitizable but is *dynamically sensitizable*. For example, again in Figure 4.6, if a and b are both switched from 1 to 0 with $c = 0$, node g switches from 1 to 0 at time 9. This implies that the path of length 9 is dynamically sensitizable. With dynamic sensitization, the propagation of input events throughout the logic is considered, and it is simply insured that side inputs are at their sensitizing value at the instant the transition passes through the gate. Note also that the latest rising delay of this circuit is 6 since is it is not possible for the path of length 9 to cause a rising transition at node g. This emphasizes the importance of also considering both rising and falling delays separately.

Within the past two years, many approaches to timing analysis that allow for dynamically sensitizable paths have been proposed [76, 21, 67, 16, 60, 19, 91, 34, 20, 54]. In general, these strategies can be classified by the assumed delay "mode". Some timing analysis work assumes the so-called *floating operational mode* delay model, as defined in [15], in which the starting state of all nodes in the combinational logic circuit is assumed to be unknown. Examples of algorithms using this model are found in [67, 15, 19]. These algorithms seek to produce a single final input vector that would possibly cause the latest (or earliest) output transition. The shortcoming of this approach is due to the fact that it is not verified whether the the internal nodes of the circuit can be put into the required beginning state. However, this approach does appear to be have advantages in terms of algorithmic complexity. The second "mode" is called the *transitional operational mode* as defined in [15] and algorithms using this model (for example [20] and [54]) seek to produce a two-vector input sequence that

Table 4.2 Sensitization Example Delay Comparison for Node g

Delay Computation Method	rising delay	falling delay
Longest Physical Path	9	9
Longest Statically Sensitizable Path	6	6
Longest Dynamically Sensitizable Path	6	9
Latest True Circuit Transition	6	8

first places the circuit in the desired setup state and then produces the desired early or late transition. The algorithms described these two chapters operate under the transitional mode in order to achieve accurate delay detection.

Previous transition mode algorithms as described above can accurately report the maximum and minimum delay values if the maximum and minimum delay of the circuit is produced by a single transition down a single path with all side inputs to an element constant at the time that the signal passes through the element. However, as it was shown in the previous section, the longest and shortest gate delays of CMOS gates can occur when multiple inputs are changing.

Revisiting the example of Figure 4.6, it was shown that the longest dynamically sensitizable path of the circuit was of length 9. This delay inherently assumes that input condition causing the worst case latest gate delay is satisfied for all gates. For each gate type shown in Table 4.1, the worst case delays occur when only a single input of a gate is changing. However, it is clear that the delay of 9 cannot be produced in which all gates along the path have only one input changing at any given time. In fact, in order for this path to produce any output transition at all, inputs a and b must both change values together and the worst case falling time of node g is 8. The rising and falling delays of node g found by the various methods of delay calculation are summarized in Table 4.2. Determining the sensitization of multiple simultaneous transitions is also important in calculating short paths since the shortest delay may only be possible when multiple simultaneous transitions can occur.

For algorithms in which the assumption of constant side inputs is made, two possibilities exist for the selection of gate delays. First, actual delays can be used in which case, if multiple input transitions produce worst case delay, this may not report worst case overall delay of the circuit. Second, as demonstrated in the example, true worst case delays can be used, whether one transition or multiple input transitions actually cause the output transition to occur. This approach is guaranteed to report a conservative bound on the critical paths but, as shown in the example, this delay may not be actually realizable depending on whether the logic and delay structure of the circuit will allow the multiple transitions to occur simultaneous at particular gates.

Therefore, in order to find the true earliest and latest events, it must be determined whether multiple events can occur simultaneously. In general, this involves the sensitization of multiple paths and determining if those paths can be sensitized simultaneously. To avoid the confusion of the terminology of path sensitization since actually multiple paths will be involved, determining if a potential event can occur at a particular circuit node will be referred to as *event sensitization* and the sensitization problem that has been identified will be referred to as *Dynamic Multiple Event Sensitization.*

4.2.4 Algorithm Strategies

There are two basic algorithmic approaches that have been used in previous algorithms for calculating critical delays and eliminating false paths. The first is a *path oriented* approach in which the circuit is modeled as a graph and paths of the graph are checked for sensitizability under various sensitization conditions. Most algorithms using the path oriented approach either generate a set of potential critical full paths and then check the sensitization of each of these paths one by one to determine the longest (or shortest) sensitizable path or check sensitization during the generation of paths and only keep those partial paths that can be sensitized. In most cases the graph modeling of circuit delays consists of a set of nodes corresponding to circuit nodes and connecting weighted edges corresponding to delays from input to output terminals of circuit elements and delays along interconnect between circuit elements. This approach can be easily extended to handle complex gate structures [76] and rising vs. falling delays. However, using a conventional path oriented approach to model circuit element delays that are dependent on whether multiple simultaneous transitions can occur would potentially cause an explosion in the number of paths in the graph.

The second approach used by current algorithms to detect false path is an *algebraic* approach. These methods typically define a boolean algebra that is able to capture both logic and timing aspects of the circuit. Examples of this approach are the Timing Boolean Algebra of [34] and [54] and the time-subscripted boolean variables of [20]. Using this method, boolean logic functions are generated that capture the sensitizability of particular circuit delays. Thus the false path detection problem is transformed into a boolean satisfiability problem. Encouraging results have been shown using this approach with simple gate models, but there is a potential for an explosion in the number of terms in the boolean functions when more accurate delay models are used in which gate delays can take on many different values depending on different input conditions.

The algorithms presented in this and the next chapters take a different approach to the sensitization problem. These methods are based on the propagation of possible circuit input events through the combinational logic. Generated events are eliminated from consideration whenever it is determined that the event cannot occur due to circuit logic constraints. In the first straightforward algorithm presented in this chapter, logical constraints for each event are propagated forward with the event as the circuit is processed. In the more efficient algorithm of Chapter 5, constraints are stored in an efficient data structure and the data structure is searched to eliminate false responses from consideration.

4.3 COMPLEXITY OF PROBLEM

In general, the problem of determining the earliest and latest output events considering simultaneous transitions and input order dependent gate delays is a significantly more complex problem than the conventional longest/shortest path problem and detecting false paths. This is primarily due to the fact that dynamic sensitization of many paths must be done concurrently to ascertain whether events can indeed propagate to the inputs of a particular gate and arrive simultaneously. For example, in order to detect the true minimum rising delay of the NAND gate in Figure 4.7, it will be necessary to determine the falling transitions from logic blocks A and B and whether two transitions can occur simultaneously. The combination of events causing the minimum rising event at the output of the NAND gate could be an earliest falling event from block A in combination with a steady high signal at the output of block B or it could be a combination of two simultaneous falling events from blocks A and B. Note that, because of the difference in delay associated with simultaneous events, two simultaneous falling events may cause the earliest rising event at the output of the NAND gate even if the two falling events are not the earliest falling events in blocks A and B.

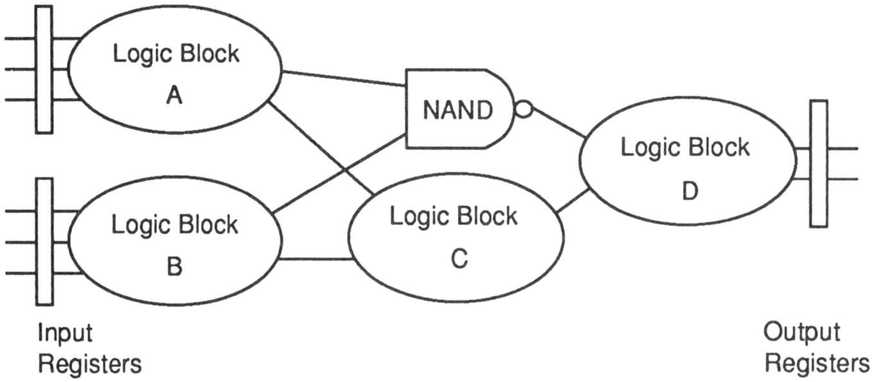

Figure 4.7 Combinational logic stage

McGeer [66] presents an overview of the complexity of several fundamental sensitization problems. In particular, it is shown that the problem of determining the longest statically sensitizable path and the problem of determining the longest dynamically sensitizable path are both NP-complete. These results are obtained from a succession of problem transformations. First the conjunctive normal form satisfiability problem (SAT), which is known to be in NP, is reduced to the problem of determining if the longest path in a circuit graph is sensitizable. This proves that the longest path sensitizability problem is NP-hard. Since this problem is shown to be a special case of the general problems of static and dynamic sensitization, these problems are also NP-hard. With the observation that a proposed solution to the sensitization decision problems can be easily verified in polynominal time, these problem are also NP-complete. With this basis, the following theorem can be stated concerning the sensitization problem proposed in this chapter.

Theorem 4.1 *Dynamic multiple event sensitization is NP-complete.*

Proof: In order to prove that this problem is NP-hard, it suffices to prove that this problem is at least as "hard" as any problem in NP. Thus, the proof of this theorem follows immediately from the fact that the longest statically sensitizable path problem in NP-complete since obviously if the latest event that is possible at the primary circuit outputs is at the output of a statically sensitizable path, this event and thus this path will be identified by a solution to the proposed sensitization problem. Thus any solution to the proposed problem will also solve the longest statically sensitizable path problem. Therefore, the proposed problem is NP-hard.

In addition, given a possible two-vector solution to the proposed problem formulated as a decision problem, it is obvious that the solution can be verified in polynomial time by a quick timing simulation with the proposed circuit inputs. Thus, the proposed problem is NP-complete. □

Since, the stated problem is in NP, there is a strong probability that any solution to the problem will have some component of its run-time that is exponential. Therefore, the goal of this work is to find a solution for the problem that is first accurate and then practical for as many circuits as possible.

4.4 NOTATION AND MODEL

4.4.1 Definitions

Based on the motivation presented in the previous section, the goal of this work is to detect both earliest and latest output events of a combinational logic block and to determine the input vectors that produce those events. To maintain accuracy, only events that can be produced by input vectors will be considered, and the data dependent effects of logic gate delays and the effect of input ordering on gate delay will be considered. In this section, definitions and notations that will be used to describe the algorithms will be presented.

The proposed analysis process is based on timing of *events* that occur a circuit nodes within the combinational logic. Events are classified as either *transition events* (circuit node rising or falling at a particular time) or *stable events* (circuit node not changing). Associated with each transition event is an *event time* representing the time at which the node changes value. Transition events are considered to be *simultaneous events* if their event times are within some constant time δ of each other. A *response waveform* at the output of a logic gate is specified by the sequence of transition events or the steady event resulting from a given excitation of the inputs to the gate. Associated with each node response waveform is a *response constraint* that describes the state of the inputs to the gate causing the response waveform to occur. The term *response* will be used to refer to the response constraint and response waveform collectively.

4.4.2 Transition Algebra

At any given instant in time, a node x can be in one of four states – high (1), low (0), rising (↑), or falling (↓). This leads to the definition of a four-valued Boolean algebra called Transition Algebra. The lattice for this algebra is shown in Figure 4.8 along with definitions of the conjunction ($\wedge_4$) and disjunction ($\vee_4$) operators. This algebra is used to compute output response waveforms based on input response waveforms at each gate. For example, the conjunction operator corresponds to the function of an AND gate and the disjunction operator corresponds to the function of an OR gate. Operators can be similarly defined for additional logic structures such as NAND, NOR, NOT, etc.

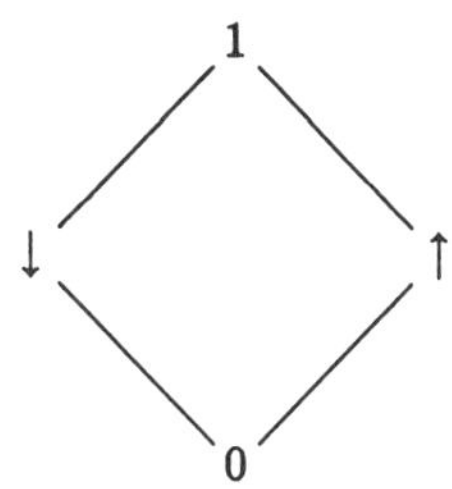

a. Lattice

$\wedge_4$	0	↓	↑	1
0	0	0	0	0
↓	0	↓	0	↓
↑	0	0	↑	↑
1	0	↓	↑	1

b. Conjunction

$\vee_4$	0	↓	↑	1
0	0	↓	↑	1
↓	↓	↓	1	1
↑	↑	1	↑	1
1	1	1	1	1

c. Disjunction

Figure 4.8 Four Valued Transition Algebra

To facilitate the manipulation of logic expressions involving this algebra (in particular, for manipulating response constraints), the following set of functions is defined to represent the four-valued system.

$x\uparrow$: true when x is rising
$x\downarrow$: true when x is falling
x : true when x is high
$\overline{x}$: true when x is low.

Table 4.3 Example NAND Event Table

	Input constraint	Event time	Output event
Transition Events	$ab\uparrow$	640ps	$\downarrow$
	$a\uparrow b$	660ps	$\downarrow$
	$a\uparrow b\uparrow$	820ps	$\downarrow$
	$ab\downarrow$	880ps	$\uparrow$
	$a\downarrow b$	780ps	$\uparrow$
	$a\downarrow b\downarrow$	500ps	$\uparrow$
Stable Events	$\overline{a}\overline{b} + \overline{a}b + a\overline{b} + \overline{a}b\downarrow + \overline{a}b\uparrow +$ $a\uparrow\overline{b} + a\downarrow\overline{b} + a\uparrow b\downarrow + a\downarrow b\uparrow$	$-\infty$	1
	ab	$-\infty$	0

These functions can be manipulated using standard two-valued boolean operators ($\cdot$ and $+$).

For example, consider the NAND gate shown in Figure 4.1. There are three possible responses with falling events at the output of the gate. That is, output c will fall when either b is rising with a steady at 1 or a is rising with b steady at 1 or a and b are both rising. The response constraints for these responses are $ab\uparrow$, $a\uparrow b$, and $a\uparrow b\uparrow$ respectively[1]. The response waveforms for these responses each contain a single falling transition event at the respective delay times.

4.4.3 Gate Delay Model

In order to accurately model gate delays, a gate model is presented that takes into consideration the effects of gate delay differences due to rising vs. falling signals and due to simultaneous vs. single input events. This gate model includes the functional operation of the gate as specified by the transition algebra operators as well as delay information between input events and resultant output events. This delay model is specified as an *event table* associated with each logic gate. This table contains entries mapping each possible input event combination to a resultant output event. For example, referring to the NAND circuit, a typical event table is shown in Table 4.3. This table is based on circuit level simulation (Figure 4.2) using nominal 2μm model parameters with all transistor widths and lengths minimum size as shown in Figure 4.4.

[1] $ab \equiv a \cdot b$

It can be seen from this table that output event times indeed vary by almost a factor of two based on relative arrival times of inputs or by approximately ±25% about the average delay of 713 ps. This clearly indicates the need for consideration of data dependent gate delays and delays dependent on input ordering.

Note that this model can be extended to multiple input gates and to different types of gates. Also, although the event times are specified here as single values, these delay could be expressed as functions of parameters such as rise time/fall time of the transition, the exact difference in arrival time of "simultaneous" transitions, etc. These extensions are discussed in more detail in Section 5.8 of Chapter 5.

4.4.4 Gate Response Representation

In performing timing analysis of a circuit, the fundamental object of interest is the response at the output of particular gates due to events at their inputs. To represent each different response at a node of the circuit, an ordered pair $(\mathcal{C},\mathcal{W})$ is used. Where $\mathcal{C}$ is the response constraint based on primary inputs that dictate the nodal response and $\mathcal{W}$ is the nodal response waveform given by a list of transition events or by a steady event. Note that allowing response waveforms to consist of multiple transition events allows for the possibility of glitches on internal circuit nodes and their potential propagation through the logic block.

4.4.5 Output Response Calculation

The process of calculating output responses for a particular logic gate involves calculation of the output response waveform and determination of the output response constraint. The output response constraint is determined by the logical AND of the input response constraints. For example, Figure 4.9 shows the calculation of an output response for a NAND gate given the specified input responses. In this case, the input responses at node a is shown to be dependent on primary circuit inputs $i1$ and $i2$ and node b is dependent on $i1$ and $i3$. (Primary circuit inputs $i1$, $i2$, and $i3$ are not shown in the figure.) The input response constraints are shown to be $i1{\downarrow}i2$ for the response at node a and $i1{\downarrow}\overline{i3}$ at node b. Thus, the output response constraint will be $i1{\downarrow}i2 \cdot i1{\downarrow}\overline{i3} = i1{\downarrow}i2\overline{i3}$.

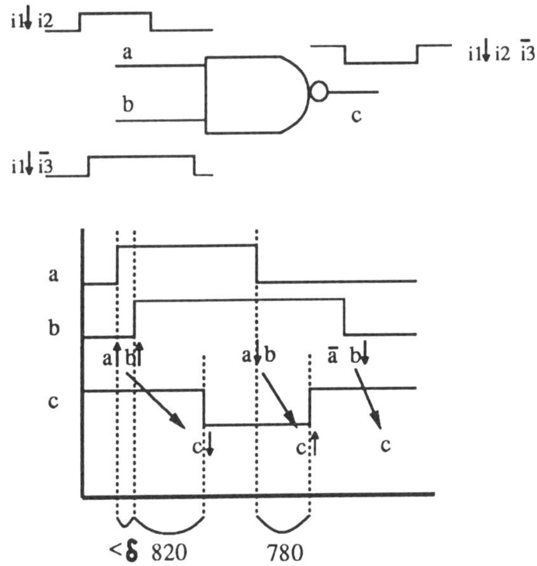

Figure 4.9 Calculation of Output Responses

To calculate the output response waveform, events of the input waveforms are examined and resultant output events are calculated using the gate event table. The sequence of output events forms the output response waveform. Figure 4.9 shows the calculation of the output response waveform for the specified input waveforms. The event times for this example are calculated from the event table of Table 4.3. Note that the first events of the input waveforms are considered to be simultaneous since the difference in their event times is less than δ. Thus, the delay of the first output event is 820. The second transition event at node a combines with a stable event at node b to produce a rising transition at c with a delay of 780. The second transition event at node b combines with a stable low event at node a and does not produce a transition event at the output. Note that this procedure does not preclude the use of more accurate delay models whose delay depends on the exact skew between input transitions, output loading, or input slew rates.

4.5 BASIC ALGORITHM DEVELOPMENT

This section presents a straightforward solution to the problem. Although this approach is computationally expensive and thus only practical for small circuits, it does demonstrate some important concepts and lays the foundation for the description of the more efficient algorithm developed in the next chapter.

The basic idea for this solution is to make a single pass forward through the gates of the logic block by levels determining output responses of each gate based on all possible input response combinations. Therefore, associated with each circuit node will be a ***response list*** containing possible nodal responses. As described above, each gate output response will contain an output response constraint based on primary inputs of the logic block and an output response waveform of the output events caused by the associated input excitation. The response list at primary input nodes implicitly contain four responses - steady high, steady low, rising, and falling. For example, the default response list for a primary input node a is $(a,\uparrow-\infty)$, $(\overline{a},\downarrow-\infty)$, $(a\uparrow,\uparrow0)$, $(a\downarrow,\downarrow0)$.

When combining responses at the input of a particular gate, it is quite possible that some combinations of input responses cannot occur together due to the previous logic. In this case, the input responses are said to be *incompatible.* Compatibility is determined using the response constraints as defined below.

Definition 4.1 *Two responses,* $\mathcal{R}_1 = (\mathcal{C}_1, \mathcal{W}_1)$ *and* $\mathcal{R}_2 = (\mathcal{C}_2, \mathcal{W}_2)$*, are said to be* **compatible** *if and only if* $\mathcal{C}_1 \cdot \mathcal{C}_2 \neq 0$.

Thus, compatibility is determined as output responses are calculated and incompatible responses are not added to the output response list. Algorithm 4.1 provides an overview of the entire timing analysis procedure.

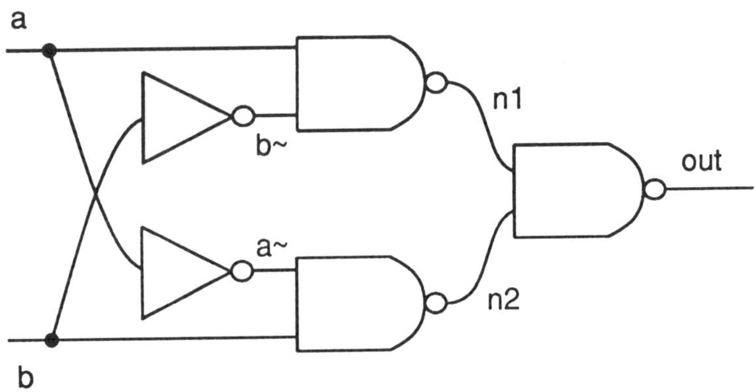

Figure 4.10 XOR Circuit Diagram

Algorithm 4.1

---

levelize the gates in the combinational logic block
put gates into queue by levels starting at inputs
do while queue not empty (forward pass)
  take gate from queue (dequeue by levels)
  for each combination of input gate responses $(\mathcal{C}_1, \mathcal{W}_1)$,$(\mathcal{C}_2, \mathcal{W}_2)$
    calculate output response constraint $(\mathcal{C}_o = \mathcal{C}_1 \cdot \mathcal{C}_2)$
    if, input responses are compatible (ie. $\mathcal{C}_o \neq 0$)
      combine 2 response waveforms to produce output $(\mathcal{W}_o)$
      output response is $(\mathcal{C}_o, \mathcal{W}_o)$
      add $(\mathcal{C}_o, \mathcal{W}_o)$ to response list of gate output
    end if
  end for
end do

To illustrate the operation of this algorithm, response tables for all nodes of the XOR circuit of Figure 4.10 are shown in Table 4.4 using the gate model of Table 4.3 (for simplicity, inverter delays are assumed to be 0). Note that incompatible responses such as $n1\!\downarrow$ at 640 and $n2\!\downarrow$ at 640 $(ab\!\downarrow \cdot \overline{a}b\!\uparrow = 0)$ do not produce any output response at node *out*.

Table 4.4 XOR Node Response Tables

XOR Circuit Node Response			
Node **a**	Node **n1**	Node **n2**	Node **out**
$(a, \ \uparrow -\infty)$	$(a\, b, \ \uparrow -\infty)$	$(a\, b, \ \uparrow -\infty)$	$(a\, b, \ \downarrow -\infty)$
$(\bar{a}, \ \downarrow -\infty)$	$(a\, \bar{b}, \ \downarrow -\infty)$	$(a\, \bar{b}, \ \uparrow -\infty)$	$(a\, \bar{b}, \ \uparrow -\infty)$
$(a\uparrow, \ \uparrow 0)$	$(a\, b\uparrow, \ \uparrow 880)$	$(a\, b\uparrow, \ \uparrow -\infty)$	$(a\, b\uparrow, \ \downarrow 1540)$
$(a\downarrow, \ \downarrow 0)$	$(a\, b\downarrow, \ \downarrow 640)$	$(a\, b\downarrow, \ \uparrow -\infty)$	$(a\, b\downarrow, \ \uparrow 1420)$
	$(\bar{a}\, b, \ \uparrow -\infty)$	$(\bar{a}\, b, \ \downarrow -\infty)$	$(\bar{a}\, b, \ \uparrow -\infty)$
	$(\bar{a}\, \bar{b}, \ \uparrow -\infty)$	$(\bar{a}\, \bar{b}, \ \uparrow -\infty)$	$(\bar{a}\, \bar{b}, \ \downarrow -\infty)$
	$(\bar{a}\, b\uparrow, \ \uparrow -\infty)$	$(\bar{a}\, b\uparrow, \ \downarrow 640)$	$(\bar{a}\, b\uparrow, \ \uparrow 1520)$
	$(\bar{a}\, b\downarrow, \ \uparrow -\infty)$	$(\bar{a}\, b\downarrow, \ \uparrow 880)$	$(\bar{a}\, b\downarrow, \ \uparrow 1520)$
Node **b**	$(a\uparrow\, b, \ \uparrow -\infty)$	$(a\uparrow\, b, \ \uparrow 780)$	$(a\uparrow\, b, \ \downarrow 1420)$
$(b, \ \uparrow -\infty)$	$(a\uparrow\, \bar{b}, \ \downarrow 660)$	$(a\uparrow\, \bar{b}, \ \uparrow -\infty)$	$(a\uparrow\, \bar{b}, \ \uparrow 1440)$
$(\bar{b}, \ \downarrow -\infty)$	$(a\uparrow\, b\uparrow, \ \uparrow -\infty)$	$(a\uparrow\, b\uparrow, \ \uparrow -\infty)$	$(a\uparrow\, b\uparrow, \ \downarrow -\infty)$
$(b\uparrow, \ \uparrow 0)$	$(a\uparrow\, b\downarrow, \ \downarrow 820)$	$(a\uparrow\, b\downarrow, \ \uparrow 500)$	$(a\uparrow\, b\downarrow, \ \downarrow 1140\ 1600)$
$(b\downarrow, \ \downarrow 0)$	$(a\downarrow\, b, \ \uparrow -\infty)$	$(a\downarrow\, b, \ \downarrow 660)$	$(a\downarrow\, b, \ \uparrow 1540)$
	$(a\downarrow\, \bar{b}, \ \uparrow 780)$	$(a\downarrow\, \bar{b}, \ \uparrow -\infty)$	$(a\downarrow\, \bar{b}, \ \downarrow 1440)$
	$(a\downarrow\, b\uparrow, \ \uparrow 500)$	$(a\downarrow\, b\uparrow, \ \downarrow 820)$	$(a\downarrow\, b\uparrow, \ \downarrow 1160\ 1700)$
	$(a\downarrow\, b\downarrow, \ \uparrow -\infty)$	$(a\downarrow\, b\downarrow, \ \uparrow -\infty)$	$(a\downarrow\, b\downarrow, \ \downarrow -\infty)$

4.6 CONCLUSION

This chapter has presented a framework for exact timing analysis to determine earliest and latest output events of a combinational logic block. This methodology is motivated by the need for accurate timing analysis in high speed CMOS systems and allows for gate models which include the effect of gate delay variation due to simultaneous vs. single events at gate inputs and the effect of input terminal ordering. Notation, models, and a basic algorithm have been presented leading toward the discussion of the full solution in the next chapter.

5

EXACT TIMING ANALYSIS: ALGORITHM

5.1 INTRODUCTION

The main problem in using the analysis strategy of the previous chapter is the exponential explosion of the space requirement for storing each of the responses and response constraints at each circuit element. This in turn causes an exponential explosion in run-time necessary to generate the responses. The approach described in this chapter relies on the fact that many of the event times of responses at a given circuit element will be identical (depending on the level of quantization of delay times). These responses with identical event times are collapsed into one response and the response constraints are stored implicitly in a Response Dependency Graph (RDG). In this way, the space requirement is drastically reduced, and the major time complexity becomes searching the RDG to determine if responses are actually possible.

This chapter begins with a discussion of the overall strategy of this method of calculating earliest and latest events along with a discussion of the structure of the RDG. The next section shows that the RDG can, many times, be searched hierarchically and outlines a method for detecting these hierarchical components. Then an algorithm is given for the verification of particular responses using this hierarchical approach and for determining the input vectors that will produce the response. Next, a prototype implementation of the algorithm is described and experimental results are shown for a set of benchmark circuits. Finally, a discussion of limitations and extensions to the outlined method is given. This section outlines some strategies for improving the runtime of the algorithm and strategies for solving larger circuits.

5.2 ALGORITHM STRATEGY

The problem, as defined in Chapter 4, is, given a circuit described as a directed acyclic graph and timing models of base level circuit elements, to determine the earliest and latest event times at all primary circuit outputs and to determine primary input vector pairs that will produce these events.

The overall strategy of this algorithm is to visit each of the circuit elements in levelized order determining the set of responses that can occur based on all combinations of responses at the inputs of the element. If there are potential false responses, all responses are checked for validity. This algorithm is shown in Algorithm 5.1.

Algorithm 5.1

Main
1. Preprocessing Step
2. *for* each circuit element (g) (in levelized order)
3. Calculate Output Responses (Update RDG)
4. *if* g is an RFI node, Verify Output Responses
5. *end for*

The preprocessing step reads the circuit and model files, generates the data structures, and detects the hierarchical subgraph components described in Section 5.5. The circuit analysis then proceeds as described above. In the following sections, each of the major steps of this algorithm is described in detail.

5.3 CALCULATION OF OUTPUT RESPONSES

The calculation of responses at each circuit element proceeds in much the same way as described in the last chapter. At a particular element, all combinations of responses at the fanin nodes are examined and the resulting response for each combination is calculated. However, instead of propagating response constraints forward with the response times, only the event times for a given

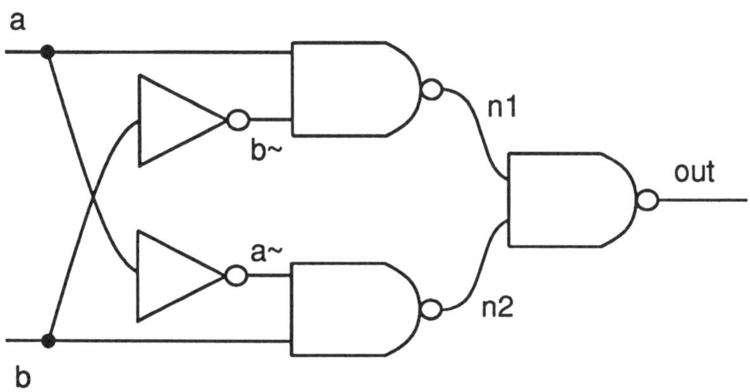

Figure 5.1 XOR Circuit Diagram

response are stored and a set of constraint pointers is created to point back to the responses of the predecessor nodes that produce the given response. Responses that have identical response times are collapsed into one response with multiple sets of constraint pointers. An example RDG for the response event 17↓ at node **out** of the XOR circuit in Figure 5.1 is shown in Figure 5.2. For this example, all delays through the inverters were assumed to be 0 and the gate model used for the NAND gates is shown below.

	Input constraint	Event time	Output event
Transition Events	$ab\uparrow$	9	↓
	$a\uparrow b$	9	↓
	$a\uparrow b\uparrow$	10	↓
	$ab\downarrow$	10	↑
	$a\downarrow b$	10	↑
	$a\downarrow b\downarrow$	8	↑
Stable Events	$\overline{a}\overline{b} + \overline{a}b + a\overline{b} + \overline{a}b\downarrow + \overline{a}b\uparrow +$ $a\uparrow\overline{b} + a\downarrow\overline{b} + a\uparrow b\downarrow + a\downarrow b\uparrow$	-∞	1
	ab	-∞	0

For comparison, the response table that would be generated by Algorithm 4.1 of Chapter 4 is shown in Table 5.1. Notice that the same timing information contained in the response table is also contained in the RDG of Figure 5.2. However, the number of responses stored and checked at each element is reduced. For example, using the strategy of the previous chapter, 16 responses

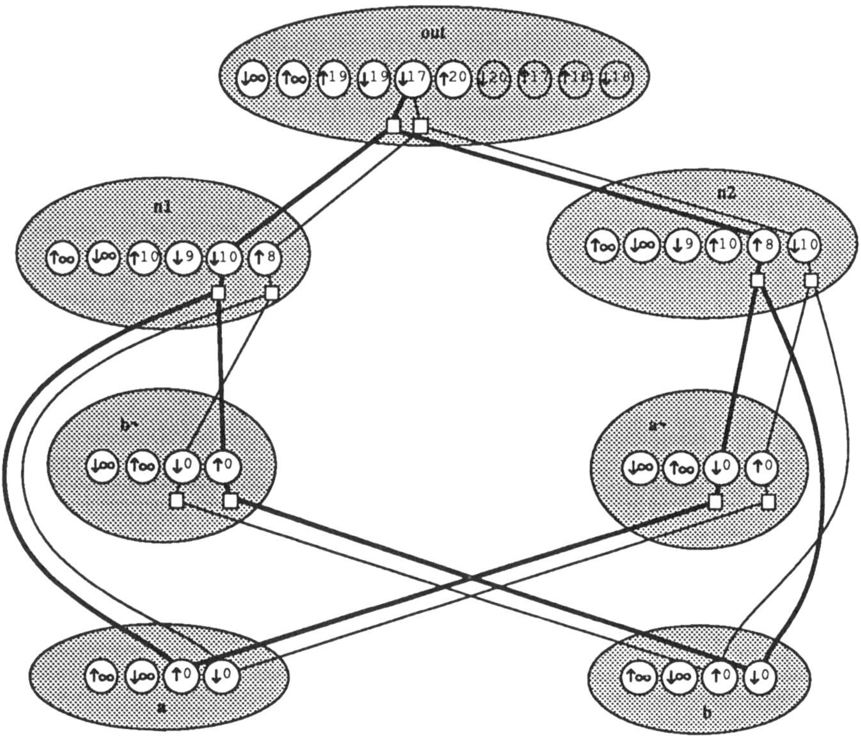

Figure 5.2 XOR RDG 1

Table 5.1 XOR Response Table

XOR Circuit Node Response			
Node **a**	Node **n1**	Node **n2**	Node **out**
$(a, \ \uparrow -\infty)$	$(a\ b, \ \uparrow -\infty)$	$(a\ b, \ \uparrow -\infty)$	$(a\ b, \ \downarrow -\infty)$
$(\bar{a}, \ \downarrow -\infty)$	$(a\ \bar{b}, \ \downarrow -\infty)$	$(a\ \bar{b}, \ \uparrow -\infty)$	$(a\ \bar{b}, \ \uparrow -\infty)$
$(a\uparrow, \ \uparrow 0)$	$(a\ b\uparrow, \ \uparrow 10)$	$(a\ b\uparrow, \ \uparrow -\infty)$	$(a\ b\uparrow, \ \downarrow 19)$
$(a\downarrow, \ \downarrow 0)$	$(a\ b\downarrow, \ \downarrow 9)$	$(a\ b\downarrow, \ \uparrow -\infty)$	$(a\ b\downarrow, \ \uparrow 19)$
	$(\bar{a}\ b, \ \uparrow -\infty)$	$(\bar{a}\ b, \ \downarrow -\infty)$	$(\bar{a}\ b, \ \uparrow -\infty)$
	$(\bar{a}\ \bar{b}, \ \uparrow -\infty)$	$(\bar{a}\ \bar{b}, \ \uparrow -\infty)$	$(\bar{a}\ \bar{b}, \ \downarrow -\infty)$
	$(\bar{a}\ b\uparrow, \ \uparrow -\infty)$	$(\bar{a}\ b\uparrow, \ \downarrow 9)$	$(\bar{a}\ b\uparrow, \ \uparrow 19)$
	$(\bar{a}\ b\downarrow, \ \uparrow -\infty)$	$(\bar{a}\ b\downarrow, \ \uparrow 10)$	$(\bar{a}\ b\downarrow, \ \uparrow 19)$
Node **b**	$(a\uparrow\ b, \ \uparrow -\infty)$	$(a\uparrow\ b, \ \uparrow 10)$	$(a\uparrow\ b, \ \downarrow 19)$
$(b, \ \uparrow -\infty)$	$(a\uparrow\ \bar{b}, \ \downarrow 9)$	$(a\uparrow\ \bar{b}, \ \uparrow -\infty)$	$(a\uparrow\ \bar{b}, \ \uparrow 19)$
$(\bar{b}, \ \downarrow -\infty)$	$(a\uparrow\ b\uparrow, \ \uparrow -\infty)$	$(a\uparrow\ b\uparrow, \ \uparrow -\infty)$	$(a\uparrow\ b\uparrow, \ \downarrow -\infty)$
$(b\uparrow, \ \uparrow 0)$	$(a\uparrow\ b\downarrow, \ \downarrow 10)$	$(a\uparrow\ b\downarrow, \ \uparrow 8)$	$(a\uparrow\ b\downarrow, \ \downarrow 17\ 20)$
$(b\downarrow, \ \downarrow 0)$	$(a\downarrow\ b, \ \uparrow -\infty)$	$(a\downarrow\ b, \ \downarrow 9)$	$(a\downarrow\ b, \ \uparrow 19)$
	$(a\downarrow\ \bar{b}, \ \uparrow 10)$	$(a\downarrow\ \bar{b}, \ \uparrow -\infty)$	$(a\downarrow\ \bar{b}, \ \downarrow 19)$
	$(a\downarrow\ b\uparrow, \ \uparrow 8)$	$(a\downarrow\ b\uparrow, \ \downarrow 10)$	$(a\downarrow\ b\uparrow, \ \downarrow 17\ 20)$
	$(a\downarrow\ b\downarrow, \ \uparrow -\infty)$	$(a\downarrow\ b\downarrow, \ \uparrow -\infty)$	$(a\downarrow\ b\downarrow, \ \downarrow -\infty)$

are generated for elements **n1** and **n2**. Therefore 256 (16^2) response combinations must be checked at element **out** to produce the 16 true responses. However, using the RDG, only six separate responses are stored at **n1** and **n2**. Therefore only 36 (6^2) response combinations must be checked at element **out** to produce ten different event times. Events that cannot be produced due to logic constraints of the circuit are shaded in the diagram.

Algorithm 5.2 describes the procedure of response calculation.

Algorithm 5.2

```
Calculate Output Responses
  for each combination of input gate responses
     combine the 2 response waveforms to produce output waveform
     for each event in the waveform
        if not a duplicate event, add event to response list
        Update RDG with response
     end for
  end for
```

5.4 VERIFICATION OF OUTPUT RESPONSES

Output responses for a particular circuit element are calculated based on all combinations of responses at the inputs of the element. However, since response constraints are not propagated with response times, it is not verified whether the two input responses can, in fact, occur together (due to the same primary input vector pair). For example, a case where a generated output response cannot be produced is shown in Figure 5.3 at node **out**. In this case, response $\uparrow$18 would occur if response $\downarrow$10 at **n1** and $\downarrow$10 at **n2** occurred together. However, no single combination of primary input responses can produce these two responses. Responses, such as these, that appear in the output response list of an element but cannot be produced by primary input vector pairs are called *false responses* defined analogous to the well known concept of *false paths.*

Definition 5.1 *False Response - Output response that can occur based on responses in the response lists of the input terminals of the element but cannot be produced by any pair of primary input vectors.*

In the end, false responses should not be reported at the primary output since doing so may lead to overly pessimistic critical delays. Also, if false responses are deleted at internal nodes, the complexity of the overall algorithm is reduced since it is unnecessary to propagate these false responses to subsequent circuit

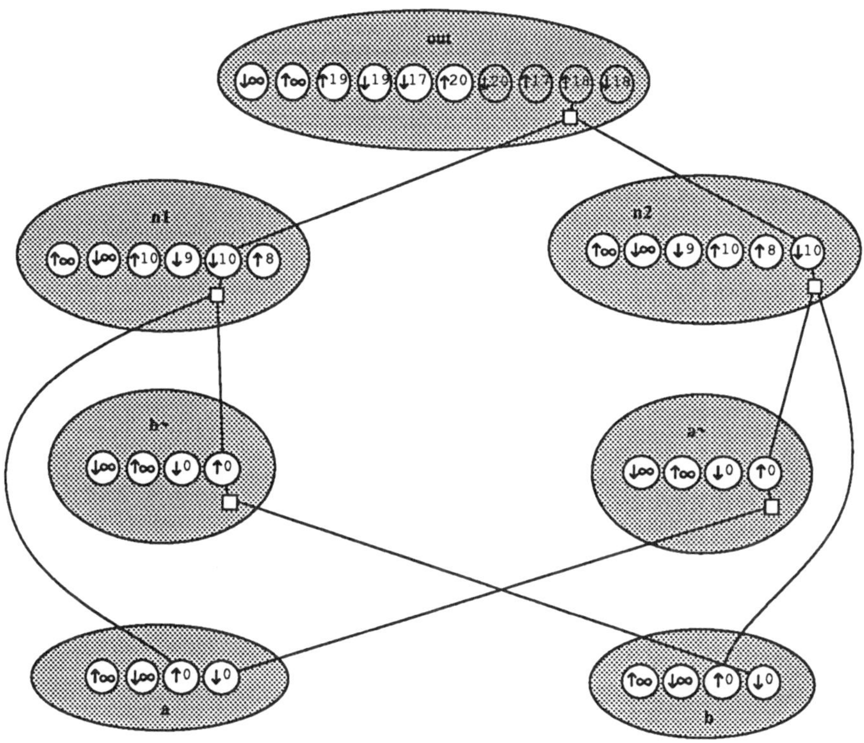

Figure 5.3 XOR RDG 2

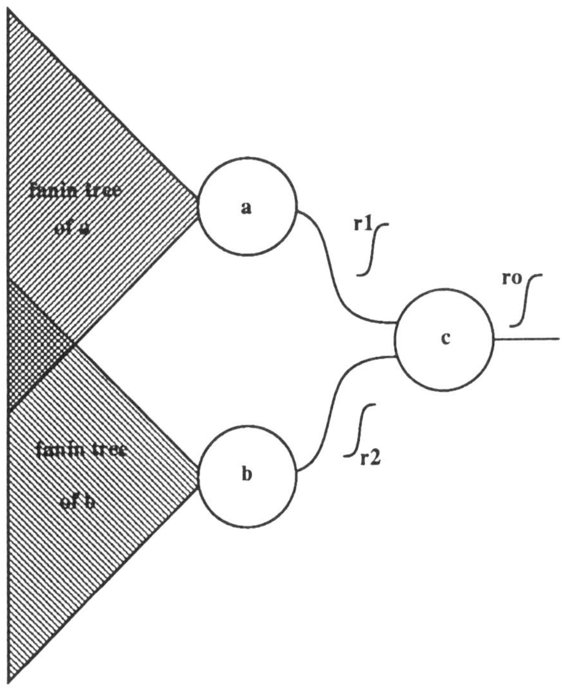

Figure 5.4 A Potential False Response

levels. Unfortunately, the detection of false responses is a computationally expensive task.

Since the goal is obviously to remove the false responses, the first task is to determine which elements can possibly have false responses. The prerequisite for the existence of a potential false response (such as **ro** at **c** in Figure 5.4) is having input responses (**r1** and **r2**) that, while possible at the previous nodes (**a** and **b**), cannot occur together due to the same primary input vector pair. This implies some common dependency in the fanin trees of the two elements. Otherwise, disjoint sets of primary input elements could produce the two responses. This dependency implies a fanout at the output of some previous level element. Therefore, for an element to have potential false responses, the element must be at the reconvergence of one or more fanout paths. This implies that the only elements at which responses must be verified are the Reconvergent Fanin (RFI) elements as shown in step 4 of Algorithm 5.1.

This fact is summarized in the following definitions and observation.

Definition 5.2 *Reconvergent Fanin Element (RFI) - A circuit element for which there are two or more disjoint paths from some element in its fanin tree to itself.*

Definition 5.3 *Reconvergent Fanout Element (RFO) - A circuit element for which there are two or more disjoint paths from itself to some element in its fanout tree.*

Observation 5.1 *The only elements that can have false responses are RFI elements.*

After identifying elements with potential false responses, the responses that are generated at these elements must be verified. In the simulation approach of the Chapter 4, the detection of false responses simply consisted of comparing response constraints (detecting incompatible responses). However, this necessitated propagating response constraints and storing each of the nodal response separately. In the search based approach of this chapter, the response constraints are not propagated explicitly with the nodal responses but are instead stored implicitly in the RDG. Therefore, the RDG must be searched to determine the validity of a particular response. The goal of this search is to find a subgraph of the RDG that contains exactly one response node and one constraint node at each circuit element. This subgraph must be entirely self consistent in that if a node is in the subgraph, all incident edges must be in the subgraph, and if an edge is in the subgraph, all incident nodes must be in the subgraph. An example of a valid subgraph is shown in Figure 5.2. The edges of this subgraph are shown in bold. As mentioned above, an example of an RDG in which no self-consistent subgraph can be found is shown in Figure 5.3.

Since the number of combinations stored by the RDG can be exponential, the search for a valid subgraph can, in worst case, be exponential in the size of the circuit. Therefore, it is important to minimize the size of the RDG that must be searched since this verification process will, in general, be the most computationally intensive task of the timing analysis process.

Fortunately, the RDGs of most RFI nodes can be decomposed hierarchically in such a way that the entire RDG need not be searched at once. This decomposition is accomplished by detecting the minimal size subgraph of the fanin tree

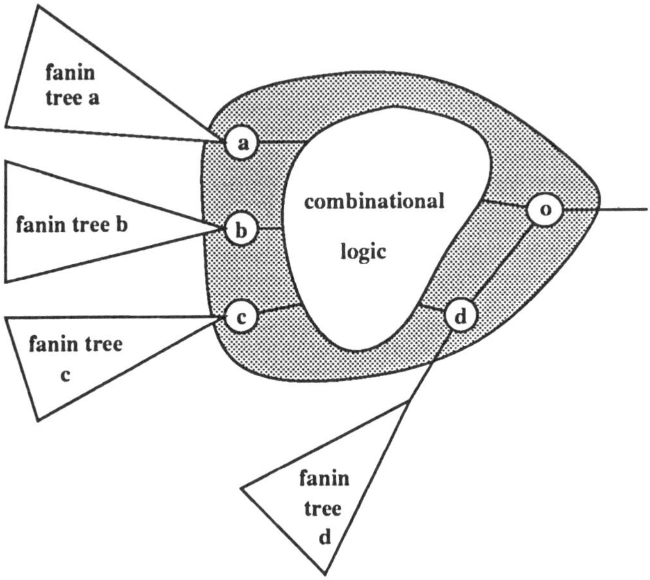

Figure 5.5 Disjoint Fanin Trees of a Component

of the circuit element under analysis such that the fanin trees of each input to the subgraph are disjoint. This is shown in Figure 5.5 for element **o**. These subgraphs are referred to as *RFI components*.

Since circuit elements are visited in levelized order, all responses at each input of an RFI component are guaranteed to be true responses. Furthermore, since the fanin trees of the inputs are disjoint, any combination of responses at the input elements can occur together due to single input vector sequence. This allows the search of the component RDG independent of the remainder of the circuit.

Figures 5.6 and 5.7 show examples of RFI components for elements **s4** and **g4_2** of a four bit adder circuit (of the type described in Chapter 7). From these figures it can be seen that the output of each component is an RFI element and the lowest level inputs to the component are a set of RFO elements. Elements are in the component if and only if they are in the fanin tree of the output and in the fanout tree of one or more of the component input nodes. The fanin trees of all inputs to each component are disjoint. Section 5.5 will give an algorithm for the detection of these components.

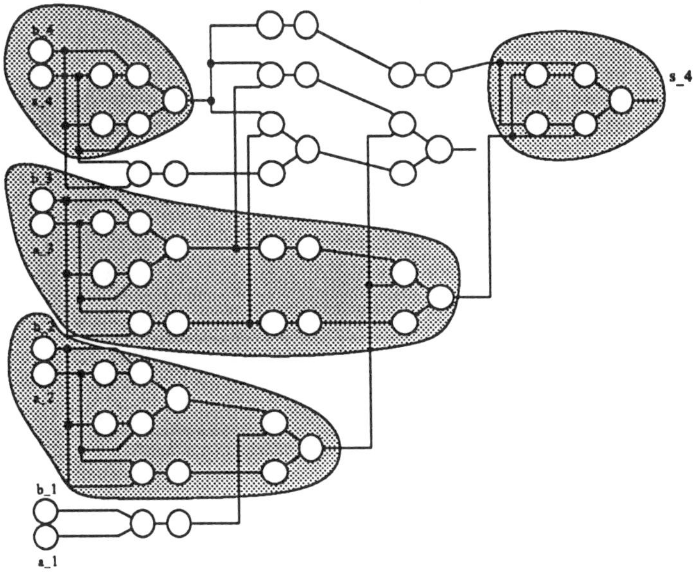

Figure 5.6 Component Example of s_4

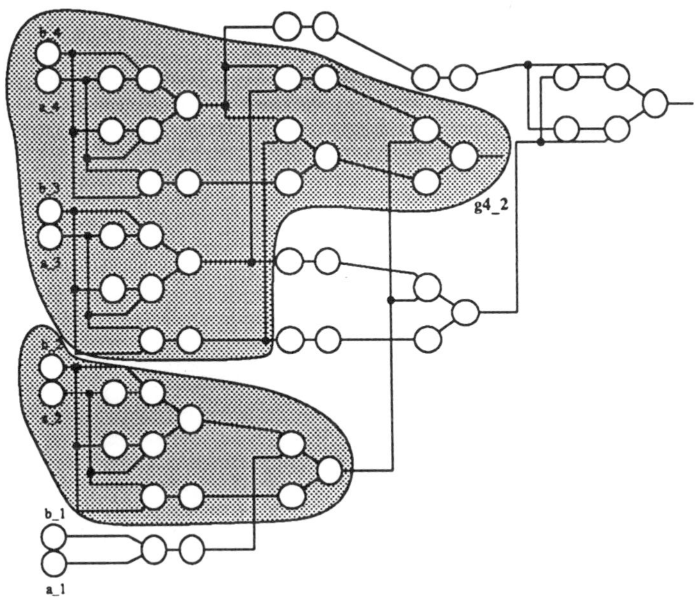

Figure 5.7 Component Example of $g4\_2$

Now, with the above definition of the component, the following theorem can be stated to show that a component RDG can indeed be searched independent of the remainder of the circuit.

Theorem 5.1 *Any response, at a node outside an RFI component that satisfies a constraint at the input of the component, can occur independent of all such responses at other component inputs.*

Proof: First, the response described in the theorem must be a true response and therefore can occur. This is because circuit elements are processed in levelized order (Algorithm 5.1) and thus the element with the response under question will have already been processed and false responses will have already been deleted.

This response must be independent of all other such responses since if there are responses at two elements that were not independent, there would be some common element in the fanin trees of the two elements. This would imply an additional RFO point and the inclusion of the RFO element inside the component. □

This theorem then allows the algorithm to pick any response at an element outside the component to satisfy a constraint and be assured that the chosen response can occur without explicitly checking. Thus, each component can be searched independently. Algorithm 5.3 outlines a method for verifying all responses at a component output circuit element.

Algorithm 5.3

```
Verify Output Responses
for each response (r) at the element (g)
   Mark all elements in component
   Initialize all responses in component
   if not SelectResponse(0)
      Delete Response - Response is not sensitizable
   end if
end for
```

Algorithm 5.4

```
SelectResponse(i)
 1.  for each response r at the ith component input
 2.     if r is consistent with other responses already chosen at
            previous component inputs
 3.        if i is the last input of component, return(TRUE)
 4.        else
 5.           if SelectResponse(i + 1), return(TRUE)
 6.           end if
 7.        end if
 8.     end if
 9.  end for
10.  return(FALSE)
```

The overall strategy as outlined in Algorithm 5.3 is to visit each of the responses at the component output circuit element (g) checking each one for validity. Steps 2 and 3 represent preprocessing steps to initialize elements of the component rooted at g. Step 4 calls the procedure SelectResponse which returns FALSE if the response is a false response. If so, step 5 deletes the response from the response list. At the end of the process, all responses in the response list are true responses.

The validity check in step 4 begins a bottom-up search of the RDG to find a set of responses at the inputs of the component that produce the desired output response **r**. This process is implemented as a recursive procedure SelectResponse(i). In the preprocessing steps of Algorithm 5.3, all responses at circuits elements in the component that can possibly cause the output response **r** are identified and marked using a depth-first search of the RDG. Note that several response nodes are, in general, identified at each circuit element. Also, in the preprocessing step, the input elements of the component are ordered starting with 0. As shown in Algorithm 5.4, a call to procedure SelectResponse(i) searches through the list of marked responses at the ith component input and attempts to find a response that is consistent with other responses chosen at previous component inputs. This consistency check involved propagating the implications of the response choice at input **i** to the other elements of the component. If the response is consistent, the implications are recorded by marking valid responses at component elements, and the procedure is recursively called on the next component input (**i+1**). If a consistent response is found at this next input, the procedure returns TRUE. Otherwise, an attempt is made to find another consistent response at the current element (**i**). If no consistent response can be found, the procedure returns FALSE.

As an example of this verification process, consider the verification of ↓17 at element **out** of Figure 5.2. Figure 5.8 shows the RDG after the initialization step (step 3) of Algorithm 5.3. Unmarked responses are shaded. The component inputs in this example are elements **a** and **b**. To begin the verification process, SelectResponse(0) (on element **a**) is called. SelectResponse selects response ↑0 at element **a**, and the implications of this selection are propagated through the component. Since there are no previously selected responses, ↑0 is consistent, and the implied responses are marked. These responses are indicated in Figure 5.8 by the bold line response nodes. SelectResponse(1) (on element **b**) is then called. The first possible response here is ↑0. However, this response is not consistent with the response chosen at element **a** since ↑0 at **b** cannot cause ↓10 at **n1** or ↑8 at **n2**. Thus, SelectResponse selects response ↓0 at **b**. This response is consistent with ↑0 at **a**. Since this is the last component input element, SelectResponse(1) returns TRUE and thus SelectResponse(0) returns TRUE indicating that ↓17 at **out** is a true response.

Notice that, in worst case, the total number of consistency checks could be the product of the number of responses at all of the component input elements and thereby exponential in the number of component inputs. However, while this inevitably limits the size of the circuits that can be analyzed by this process, two observations make the procedure workable for many cases. First, the component size does not necessarily grow with the size of the circuit. That is, many times, the component size will be manageable even for larger circuits. Second, very rarely would all combinations of responses at the component inputs need to be considered. For example, if a response at the second component input is found to be inconsistent with a response at the first component input, none of the combinations of responses involving this combination of first and second input responses will be considered. As a final note on the efficiency of this algorithm, it should be noted that these observations are circuit dependent and therefore will not hold true for all circuits. In general, this validation process is the limiting factor in terms of runtime in the analysis of larger circuits.

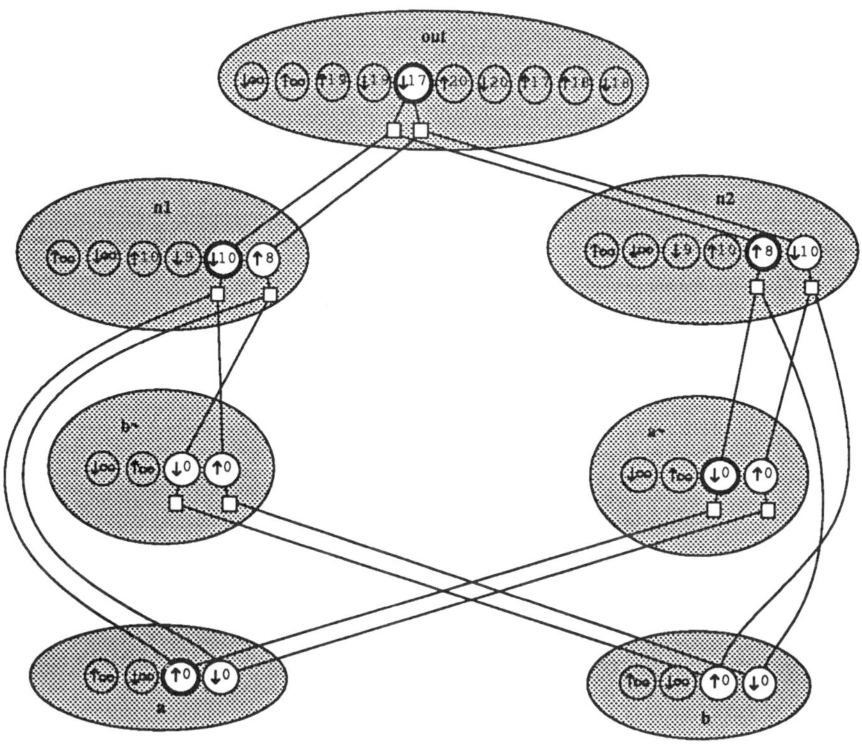

Figure 5.8 Verification of ↓ 17

5.5 DETECTION OF COMPONENTS

The previous section identified a type of self-contained component of the circuit graph in which the fanin trees of all inputs to the component are disjoint. This is used in the verification of output responses to reduce the size of the RDG that must be searched at once. This section concerns the detection of these components.

Algorithm 5.5 describes the method used to detect all RFI components for all elements in the circuit. This procedure is run in the preprocessing step of Algorithm 5.1 and a list of component input elements is generated for each RFI element in the circuit. These component input elements are used in the verification routines of Algorithms 5.3 and 5.4. Non-RFI elements need not be analyzed as shown in the previous section.

Algorithm 5.5

Find Component

1. *for* each element (g) (in levelized order)
2. Find all paths from g to all predecessor elements
3. Sort paths by source element
4. Detect all pairs of disjoint paths to all predecessor elements and add source element to component input list of g. Also add the component input list of all internal nodes of both paths to the component input list of g.
5. *end for*

This algorithm visits each element in the circuit graph with the goal of finding a set of elements in the fanin tree of the element that have disjoint fanin trees. To accomplish this, the algorithm first finds all paths from the node under question (g) to all elements in its fanin tree. After sorting the paths (for efficiency in comparison), all pairs of paths with disjoint elements except for source and sink are detected. At this point if there are two or more disjoint paths from some element to g, then g is an RFI element and there will be a component rooted in g. The inputs to that component consist of all source elements of disjoint paths as well as all component input lists of all elements in the paths. Note that the component rooted at each element in the path will have already been detected since the elements are visited in levelized order. The following theorem proves that this type of detection will results in components with the desired properties. Refer to Figures 5.6 and 5.7 for component examples.

Theorem 5.2 *Algorithm 5.5 identifies a component of the circuit graph such that all inputs have disjoint fanin trees.*

Proof: If the fanin trees were not disjoint, then there would be a least one circuit element in common such that there would be two disjoint paths from this element to some element inside the component passing through two different component inputs. However, by Algorithm 5.5, this node would also have to be in the component. Thus the theorem is proved. □

5.6 VECTOR REPORTING

After all responses have been verified at each node, the full primary input vector pair that will produce a given response may be found. Algorithm 5.6 outlines a method for finding a full input vector pair that will produce the given response. It is assumed that verification has already been done, and thus all responses at all circuit elements are true responses.

Algorithm 5.6

```
FindVector(g,r)
if element g is a primary input return(TRUE)
if element g is a RFI element
   Mark some combination of responses at each boundary node of the
         reconvergent component that produce response r at g.
   for each boundary node b and its marked response r1
      FindVector(b,r1)
   end for
else
   Mark some combination of responses at each fanin of g that
         will produce response r at g.
   for each Fanin node f and its marked response r1
      FindVector(f,r1)
   end for
end if
```

This process is basically a recursive implementation of the process that was used to verify responses. To begin, the FindVector procedure is called on the output element and the output response under question. Analysis proceeds on the component rooted at the output element and then recursively on all boundary elements of the component and the associated chosen response. Since the same process is used as in the verification process, the runtime of this process can be exponential in the number of component inputs. However, if the verification process completed, this procedure is guaranteed to complete since this is simply a repeat of analysis previously done.

5.7 IMPLEMENTATION

The algorithm described in this chapter has been implemented in a prototype timing analyzer, *XTV*. It is written in C and runs under Unix on a DECstation 3100/DECstation 5000 and also under MSDOS. Input to *XTV* is a circuit file in ISCAS85 format and output is a report of earliest and latest output event times along with vectors to excite those events for the node under question.

Table 5.2 Benchmark Circuit Statistics

Circuit	Number Gates	Number Inputs	Number Outputs
adder4	83	8	5
adder16	523	32	17
adder4.pla	71	9	5
br1	106	12	8
co14	65	14	1
dc1	42	4	7
f51m	163	8	8
in6	261	33	23
log8mod	138	7	5
luc	216	8	27
max46	165	9	1
mp2d	103	14	14
newapla	55	12	10
newtpla2	39	10	4
p82	91	5	14
rd73	141	7	3
risc	93	8	30
ryy6	31	16	1
signet	303	39	8
sqn	117	7	3
td1	14	4	2
ts10	264	22	16
vg2	108	25	8
x11dn	120	27	6
x6dn	351	38	5
xor8	35	8	1

5.7.1 Results

Table 5.2 lists statistics from a set of benchmark circuits used to test the prototype timing analyzer. Circuits *adder4* and *adder16* are 4-bit and 16-bit parallel carry adders of the type to be described in Chapter 7. These circuits consist exclusively of biased CMOS inverters and two input biased CMOS NAND gates. The remaining circuits are from [82] and have been resynthesised as multilevel circuits using at most two input standard static CMOS logic gates. All circuits are in the ISCAS85 circuit format citeiscas85.

Table 5.3 summarizes results from *XTV* on this set of benchmark circuits. *XTV* completed analysis in less than one hour on all but the four circuits with the highest number of inputs. In these circuits the response verification process on a limited number of responses was stopped after a certain number of iterations in order to reduce the runtime. This is described more in Section 5.7.2. These responses were typically in components with a large number of input elements and the responses were simply assumed to be true responses for the remainder of the analysis process. This guarantees a safe bound on the worst case delay. In all cases, however, output responses with maximum and minimum rising and falling delays were fully verified, and two-vector input pairs were found that produce these worst case responses. This indicates that the worst case delay found for these cases was exact.

Three execution times are given for each circuit. The first time (FC) is time required to detect all components of the circuit (step 1 of Algorithm 5.1). The second time (RCV) is the time required to calculate and verify every response at every circuit element in the circuit (steps 2-5 in Algorithm 5.1). The third time (FV) is the time required to find input vector pairs that produce the maximum and minimum rising and falling delays. As shown the preprocessing and postprocessing step take negligible amounts of time in most cases.

Table 5.4 shows a comparison of *XTV* results using different gate delay models. This table shows minimum rising and falling delays for the benchmark circuits using a constant delay model and the full data dependent delay model. The constant delay models use the minimum rising gate delay for all rising transitions and the minimum falling gate delay for all falling transitions. This is the delay model used in most timing analysis to date. As shown, in all cases the constant delay models produce conservative results since worst case delay models are used. However, these results can be as much as 27% different from the exact data dependent delay results. This demonstrates the need for data dependent delay models considering the effects of simultaneous input transitions and input gate ordering for accurate timing analysis.

The choice of delay models also has a major impact on the execution time as shown in Table 5.5. This table shows the runtimes for response calculation and verification along with the number of responses that were not verified due to reaching the iteration limit (see Section 5.7.2). The constant delay model is as described above in which worst case conservative delays are used for both rising and falling delays. The unit delay model uses a delay of one time unit for all delay values. As shown in the table, runtimes are drastically reduced as simpler delay models are used. For example, the runtime for *max46* is reduced over 99% when the unit delay model is used. This is due to the fact that more

Table 5.3 *XTV* Runtime Results

Circuit	Output Node	τ_{min} (ps) rise,fall	τ_{max} (ps) rise,fall	runtime (s) FC	RCV	FV
adder4	s4	1780 1780	2340 2340	0.6	0.5	0.0
adder16	s16	2560 2560	3680 3680	22.3	1781.7†	0.8
adder4.pla	out_0	3910 3450	11122 10684	0.8	24.4	0.4
br1	out_5	3200 2910	14792 14804	2.9	298.0	1.0
co14	out_0	3374 3030	11460 11230	1.0	198.1	2.4
dc1	out_4	2820 2550	5602 5410	0.2	1.0	0.0
f51m	out_1	3134 3010	10962 10524	2.7	244.2	0.6
in6	out_10	2050 1700	12980 12720	3.9	1.6h†	1.4
log8mod	out_4	3200 2910	10662 10070	2.0	75.1	0.8
luc	out_5	1860 1750	8392 8400	3.4	60.8	0.1
max46	out_0	3970 3530	14592 13770	7.5	2646.3	8.5
mp2d	out_9	2370 2140	9978 9300	0.8	0.9	0.0
newapla	out_0	900 950	7882 7918	0.4	1.4	0.0
newtpla2	out_2	1330 1310	7308 7110	0.2	0.2	0.0
p82	out_12	2050 1930	7968 7648	0.9	4.7	0.1
rd73	out_1	3380 3010	13396 12320	3.5	844.5	3.7
risc	out_9	2840 3000	6640 6900	0.5	0.9	0.0
ryy6	out_0	2770 3110	8632 8660	0.2	2.6	0.1
signet	out_2	3200 2910	13462 12564	5.9	16.6h†	3.6
sqn	out_2	3134 3080	11898 10960	1.7	76.3	0.6
td1	out_0	1690 1520	4174 4030	0.1	0.1	0.0
ts10	out_15	3270 3350	8058 7880	9.5	808.6	1.3
vg2	out_3	3850 3370	14652 14750	1.8	325.3	0.6
x11dn	out_3	3858 3810	8688 8610	2.1	329.9	0.1
x6dn	out_3	2664 2490	15202 14140	18.4	6.9h†	4.6
xor8	out_0	5690 5330	7474 7300	0.4	156.4	1.8

† Verification of some responses were stopped at 5000 iterations of SelectResponse.

FC: Find Component
RCV: Response Calculation and Verification
FV: Find Vectors for Maximum and Minimum Delays

Table 5.4 Comparison of Delay Models

Circuit	Constant Delay† τ_{min} (ps) rise,fall	Data Dependent‡ τ_{min} (ps) rise,fall	Δ% rise,fall
adder4	1560 1560	1780 1780	-12.4 -12.4
adder16	2340 2340	2560 2560	-8.6 -8.6
adder4.pla	3380 3040	3910 3450	-13.6 -11.9
br1	2480 2320	3200 2910	-22.5 -20.3
co14	3020 2600	3374 3030	-10.5 -14.2
dc1	2480 2240	2820 2550	-12.1 -12.2
f51m	2590 2240	3134 3010	-17.4 -25.6
in6	1690 1520	2050 1700	-17.6 -10.6
log8mod	2480 2320	3200 2910	-22.5 -20.3
luc	1690 1520	1860 1750	-9.1 -13.1
max46	3270 3120	3970 3530	-17.6 -11.6
mp2d	2010 1960	2370 2140	-15.2 -8.4
newapla	900 720	900 950	0.0 -24.2
newtpla2	1330 1080	1330 1310	0.0 -17.6
p82	1690 1520	2050 1930	-17.6 -21.2
rd73	3020 2600	3380 3010	-10.7 -13.6
risc	2840 2590	2840 3000	0.0 -7.4
ryy6	2770 2880	2770 3110	0.0 -7.4
signet	2480 2320	3200 2910	-22.5 -20.3
sqn	2590 2240	3134 3080	-17.4 -27.3
td1	1690 1520	1690 1520	0.0 0.0
ts10	3270 3120	3270 3350	0.0 -6.9
vg2	3490 2960	3850 3370	-9.4 -12.2
x11dn	3490 2950	3858 3810	-9.5 -22.6
x6dn	2120 1880	2664 2490	-20.4 -24.5
xor8	5070 4560	5690 5330	-10.9 -14.4

† Single delay for rising transitions/Single delay for falling transitions
‡ Delay considering input ordering and multiple simultaneous transitions

responses at a particular circuit node will have the same event transition time thus reducing the overall number of different responses at a node that must be verified. However, as was shown in Table 5.4, this reduction in runtime comes at the price of less accuracy.

Finally, in Table 5.6, results from *XTV* using data dependent delay models are compared with results from the CAzM [23] circuit level simulator from MCNC. For these tests *XTV* is run in a simulator mode given specific input transition vectors and reporting the delay of the output transition. The *adder4* circuit contains NAND gates and inverters with fanout of 1, 2, and 3. For these tests, gates with different fanouts are modeled separately. As can be seen, in all cases, the delay reported by *XTV* is within about 6% of CAzM. This difference could be decreased even further with more accurate models including the effects of input loading and relative skew between input transitions. The first four vectors are the earliest rising, earliest falling, latest rising and lastest falling vectors for output *s4* as reported by *XTV*.

5.7.2 Enhancements

In the analysis of larger circuits, it is important to have the ability to gracefully degrade the accuracy of the solution (always insuring a conservative solution) in order to keep runtimes feasible. As mentioned in the previous section, the process of searching the RDG to verify each response is generally the most time consuming portion of the algorithm and thus will limit the size of circuits that can be analyzed in a reasonable amount of time. However, the amount of time to verify particular responses can vary widely depending on the size of the component and the exact structure of the RDG. In fact in many cases, a relative few of the responses at a node will consume the majority of the verification time.

One strategy to allow for this tradeoff between accuracy and runtime, as has been alluded to earlier, is to simply assume difficult-to-verify responses are valid without fully checking validity. This strategy is implemented in *XTV* by allowing the user to specify a limit on the number of times that the procedure SelectResponse can be called in the verification of a single response. This, in effect, sets a limit on the number of recursive iterations that will be made. If the response has not been verified when the recursion limit is reached, the response is simply assumed to be valid. This insures that a conservative bound is always kept.

Table 5.5 Comparison of Delay Models: Runtimes

Circuit	Unit Delay‡		Constant Delay*		Data Dependent**	
	runtime (s)	#†	runtime (s)	#†	runtime (s)	#†
adder4	0.1	-	0.1	-	0.5	-
adder16	1.6	-	4.4	-	1781.7	22
adder4.pla	0.7	-	1.1	-	24.4	-
br1	8.0	-	20.2	-	298.0	-
co14	1.3	-	2.3	-	198.1	-
dc1	0.2	-	0.3	-	1.0	-
f51m	7.7	-	20.4	-	244.2	-
in6	104.4	2	588.3	9	1.6h	33
log8mod	3.7	-	11.1	-	75.1	-
luc	4.1	-	8.2	-	60.8	-
max46	14.0	-	90.4	-	2646.3	-
mp2d	0.2	-	0.3	-	0.9	-
newapla	0.2	-	0.3	-	1.4	-
newtpla2	0.1	-	0.1	-	0.2	-
p82	1.1	-	1.8	-	4.7	-
rd73	5.5	-	35.8	-	844.5	-
risc	0.3	-	0.4	-	0.9	-
ryy6	0.1	-	0.1	-	2.6	-
signet	272.8	3	1722.6	10	16.6h	142
sqn	3.5	-	11.3	-	76.3	-
td1	0.0	-	0.0	-	0.1	-
ts10	10.4	-	15.3	-	808.6	-
vg2	2.0	-	4.0	-	325.3	-
x11dn	1.9	-	4.6	-	329.9	-
x6dn	38.5	-	731.3	3	6.9h	15
xor8	0.1	-	0.1	-	156.4	-

† Number of responses that were stopped at 5000 iterations of SelectResponse.

‡ Delay of 1 for all transitions

* Single delay for rising transitions/Single delay for falling transitions

** Delay considering input ordering and multiple simultaneous transitions

Table 5.6 Comparison of *XTV* results with CAzM for *adder4*

Vector	*XTV*		CAzM		Δ%
	delay (ns)	time† (sec)	delay (ns)	time† (sec)	
$a4\ a3\ a2\ a1\ b4\ b3{\uparrow}\ \overline{b2}\ \overline{b1}$	3.41	0.1	3.37	25	1.2
$a4\ \overline{a3}\ \overline{a2}\ a1{\downarrow}\ b4\ b3{\downarrow}\ b2{\downarrow}\ b1{\downarrow}$	3.02	0.1	2.90	25	4.1
$a4\ a3\ \overline{a2}\ a1\ \overline{b4}\ \overline{b3}\ b2{\downarrow}\ b1$	4.08	0.1	4.21	25	-3.1
$a4\ a3\ \overline{a2}\ a1{\uparrow}\ \overline{b4}\ \overline{b3}\ b2{\uparrow}\ b1{\uparrow}$	4.17	0.1	4.22	25	-1.2
$a4\ a3\ a2{\downarrow}\ a1{\downarrow}\ \overline{b4}\ \overline{b3}\ \overline{b2}\ b1{\downarrow}$	3.57	0.1	3.58	25	-0.3
$a4\ a3\ a2{\downarrow}\ a1{\downarrow}\ b4\ \overline{b3}\ \overline{b2}\ b1{\downarrow}$	3.12	0.1	3.08	25	-1.3
$a4\ a3{\uparrow}\ a2{\uparrow}\ a1{\uparrow}\ b4\ \overline{b3}\ \overline{b2}\ b1{\uparrow}$	3.79	0.1	3.93	25	-3.6
$a4\ a3{\uparrow}\ a2{\uparrow}\ a1{\uparrow}\ \overline{b4}\ \overline{b3}\ \overline{b2}\ b1{\uparrow}$	4.10	0.1	4.20	25	-2.4
$a4{\uparrow}\ a3{\uparrow}\ a2{\uparrow}\ a1{\uparrow}\ b4{\uparrow}\ b3{\uparrow}\ b2{\uparrow}\ b1{\uparrow}$	3.48	0.1	3.29	25	5.8
$a4{\downarrow}\ a3{\downarrow}\ a2{\downarrow}\ a1{\downarrow}\ b4{\downarrow}\ b3{\downarrow}\ b2{\downarrow}\ b1{\downarrow}$	3.03	0.1	3.02	25	0.3
$\overline{a4}\ a3{\uparrow}\ a2{\uparrow}\ a1{\uparrow}\ \overline{b4}\ \overline{b3}\ \overline{b2}\ b1{\uparrow}$	3.79	0.1	3.92	25	-3.3
$\overline{a4}\ a3{\downarrow}\ a2{\downarrow}\ a1{\downarrow}\ \overline{b4}\ \overline{b3}\ \overline{b2}\ b1{\downarrow}$	3.02	0.1	2.92	25	3.4

† Run time on DECstation 3100

It is important to note that even if some responses are not fully verified and are assumed to be true, it may still be possible to verify whether the resulting worst case delay times are exact. If the FindVector procedure can produce an input vector pair for a given reported worst case output response, the response is indeed true and the bound is exact. This situation can occur when assumed-true responses at internal nodes are not needed to produce the worst case circuit output response under consideration. For the particular circuits shown in the previous section, the FindVector procedure was able to complete on the earliest and latest output responses implying that these reported worst case delay times were exact.

Of course, the FindVector procedure cannot, in general, be guaranteed to complete in a reasonable time for all output responses when some internal node responses are not fully verified. Since the FindVector procedure uses a similar approach to the verification procedure, the FindVector procedure may also have difficulty completing in a reasonable amount of time if a assumed-true response at an internal node is needed to produce the output response under consideration. Thus, it may not be possible to produce an input vector pair for some output responses. Even in these cases, though, a conservative bound on the worst case delays can be reported.

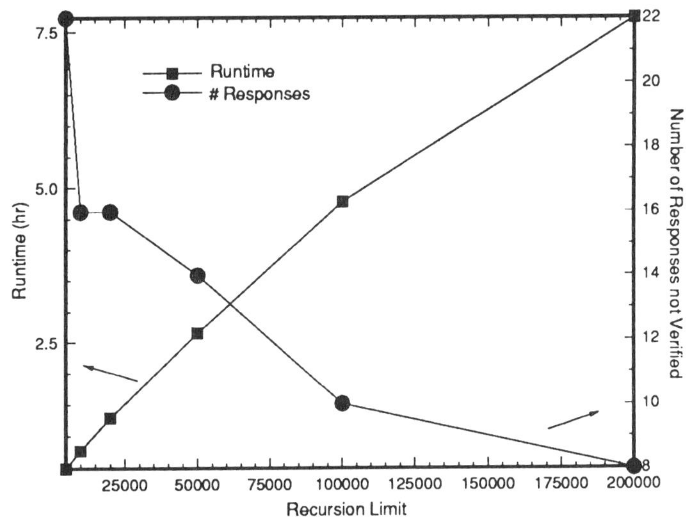

Figure 5.9 *XTV* Results on ***adder16***

Table 5.5 gives the number of responses that were not fully verified with the recursion limit set at 5000 iterations for the benchmark circuits. The number of responses that were not verified in all cases was a small percentage of the total number of responses. For example in the circuit ***signet***, only 142 out of more than 7400 (< 2%) total responses that were verified. As noted previously, input vector pairs were found for all worst case responses.

Figure 5.9 gives results of *XTV* applied to ***adder16*** with increasingly higher limits placed on the number of calls to SelectResponse. As can be seen, there is a direct tradeoff between the runtime of the algorithm and the number of responses that had to be assumed valid. For this circuit, however, the reported worst case delay times were the same for all cases, and input vector pairs were found for all worst case responses.

5.8 LIMITATIONS AND EXTENSIONS

All timing analysis systems must, to some degree, trade accuracy for speed. This section discusses areas where these tradeoffs have resulted in accuracy concessions for the presented timing analysis method. Also discussed is whether these concessions are inherent in the methodology or a result of this implementation of *XTV*.

5.8.1 Validity of Gate Models

The core of any timing analysis process is modeling the delay of the lowest level circuit elements. Throughout the discussion of the last two chapters, a simple table based gate level timing model has been assumed with each significantly different gate structure having a different model. While this table-based model is sufficient for many applications, more detailed analysis may require more accurate gate delay models. This section discusses the possibility of including more accurate gate models within the *XTV* environment.

Since the *XTV* analysis method is a based on incremental simulation of individual responses at the circuit nodes, the delay model used for gate output response calculation can actually be quite general. As shown in Step 2 of Algorithm 5.2, the basic operation of the algorithm is simulating the operation of the particular gate given specific waveforms at the inputs of the gates. In the approach described in Section 4.4.5 and implemented in *XTV*, the simple table based gate models were used to produce the gate output waveform. The analysis methodology, though, does not preclude the used of much more accurate gate models. For example, the output response could even be calculated by performing a full circuit level simulation of the gate with the specified input waveforms. While low level circuit simulation would be much too computationally expensive, gate models more accurate than the simple table based models could certainly be incorporated.

However, as more accurate delay models are used, there may also be a corresponding increase in runtime. This manifests itself in two ways in the *XTV* algorithm. First, more detailed gate models, in general, imply more time to compute output responses. Second, more accurate gate models may increase the number of different responses that must be stored at circuit nodes since event times of different responses may now be less likely to be identical. An increase in the number of different responses that must be stored will generally increase the runtime since more responses now need to be verified. In the end,

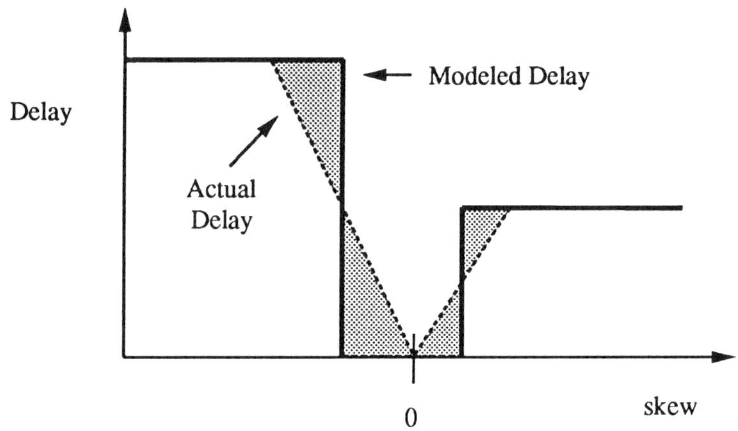

Figure 5.10 Current Gate Delay Model

there will be a tradeoff between the accuracy of the solution and the allowable runtime.

With these limitation in mind, this section discusses four possible extensions to the basic delay model considering:

- skew between multiple input transitions,
- input rise/fall time and output load,
- multiple input gates, and
- glitches.

Modeling delay due to multiple input transitions

The implemented table based gate model represents the two-input gate delay variation due to multiple input transitions with a multiple step function as shown for the rising delay of a conventional NAND gate in Figure 5.10. Transitions within time δ of each other are assumed to be simultaneous. The shaded area of the diagram shows the error associated with this modeling when delays are not exactly simultaneous.

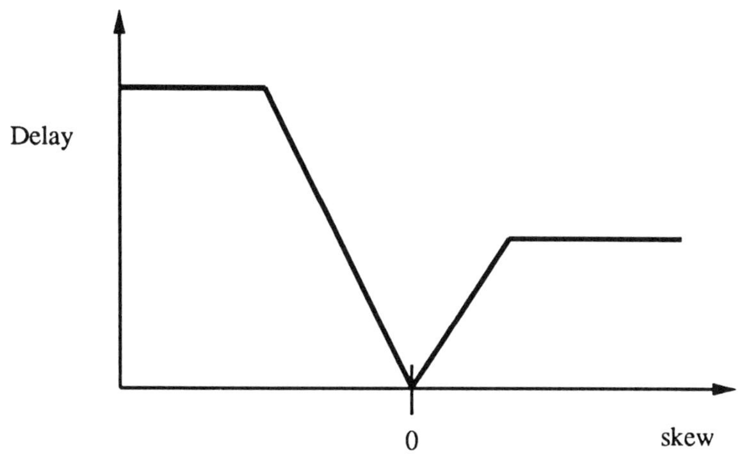

Figure 5.11 A More Accurate Gate Delay Model

An alternate and more accurate delay model is shown in Figure 5.11 in which the delay is modeled by a piecewise linear function. The model shown here is a simple piecewise linear function but, in general, more complicated functions such as piecewise cubic splines could be used to more closely match the simulated results shown in Figure 4.2. This extension would only affect Step 2 of Algorithm 5.2.

Effect of input rise/fall time and load

In the current approach, the delay of an element is independent of input rise and fall time of the input signal and the output load. Thus for accurate analysis, timing models are generated for each significantly different instance of a particular gate. For example, gates with significantly different output load would have separate gate models. To see that this is necessary, consider the effect on two input NAND gate delay due to the load and input rise time as shown in Figures 5.12 and 5.13. The test circuit configuration and transistor sizes are shown in Figure 5.14. For the output load simulations capacitors Ca and Cb were removed, and Cl was varied from 0 to 200 fF. For the input slope simulations capacitor Cl was removed, and Ca, Cb were varied together from 0 to 200 fF. Note that these results are similar to results reported in [86].

Instead of modeling different instances of gates separately, however, the effects of load and transition time could be incorporated in the gate model again

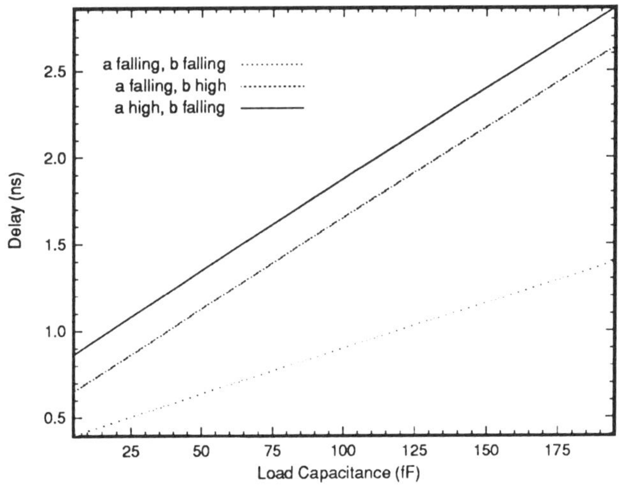

Figure 5.12 Static CMOS Delay vs. Output Load

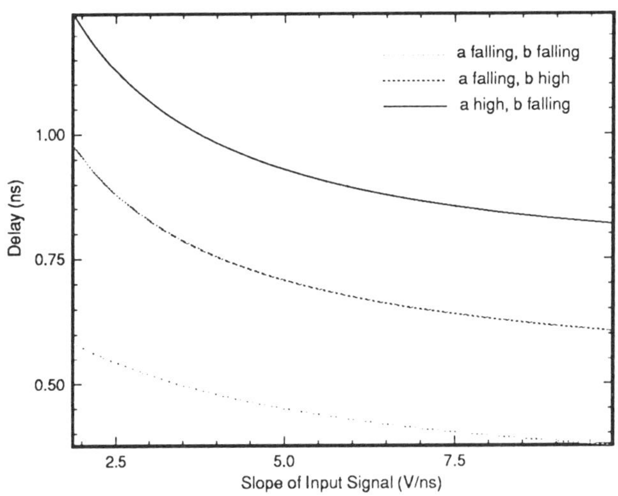

Figure 5.13 Static CMOS Delay vs. Input Slope

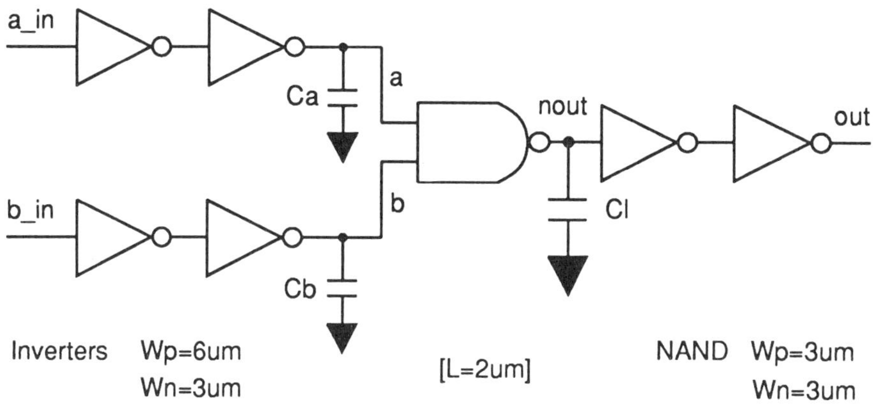

Figure 5.14 Circuit Setup for Input Slope and Output Load Simulations

through the use of piecewise linear or cubic spline models. This would require the propagation of signal transition time along with delay and would potentially increase the number of responses stored at each node since responses would now be identical only if both event delay time and event transition time are identical.

Multiple input gates

The discussion of the past two chapters has concentrated exclusively on the analysis of circuit elements with at most two inputs. Using the table based approach, the modeling of multiple input elements simply means adding more entries to the table. However, the number of entries grows exponentially with the number of input. For example, a two input gate has 16 (4^2) entries, a three input gate has 64 (4^3), a four input gate has 256 (4^4), etc. Since the specification of gate models can be automated using circuit simulation and waveform analysis[27], elements with a reasonably small number of inputs could be modeled.

In addition, the extensions described above concerning the effects of multiple transitions and the effects of loading and transition time could also be incorporated in a multiple input gate model. For example, the exact rising delay of a three input NAND gate is shown in Figure 5.15. As can be seen, this delay function can also be relatively accurately approximated by a piecewise-linear function in three dimensions. Similar conclusions can be made about the modeling of loading and input transition time effects.

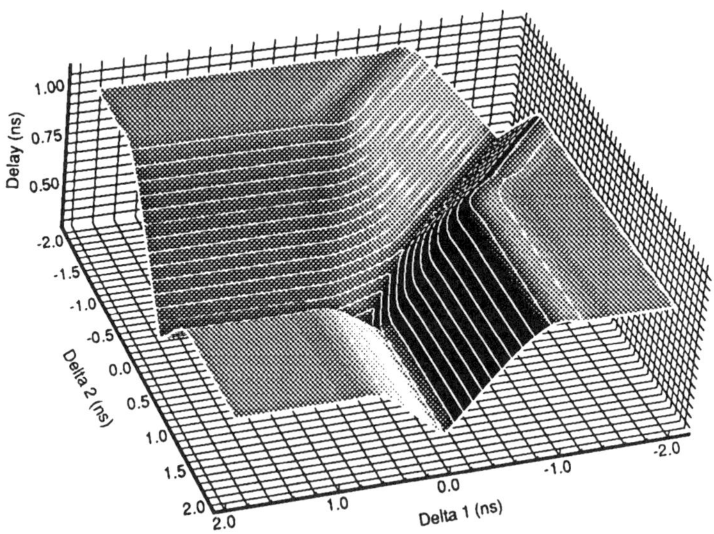

Figure 5.15 Three Input NAND Gate Rising Delay

The main effect of this extension to *XTV* would be in the storage requirements of the algorithm. Since the amount of storage needed at the output of a particular circuit element is related to the product of the number of responses at its inputs, this requirement increases exponentially as the number of inputs is increased. This is due to the fact that a pointer in the RDG must be kept for each combination of input responses. However, for a relatively small number of inputs (eg. < 5), this approach will remain feasible for most moderate sized circuits. Note that the runtime will also be correspondingly increased because of the time needed to generate these responses.

Glitches

Another inaccuracy in timing analysis occurs in the presence of glitches at circuit nodes. In the existing *XTV* strategy, a particular circuit node is considered to be either rising (full swing), falling (full swing), steady high, or steady low at any instant in time. However, when a glitch occurs at a circuit node, the response may be a less than full swing rising and falling transition as shown in Figure 5.16. *XTV* handles this case in one of two ways. First, if the two transitions are close enough to each other to cause only partial transitions, the

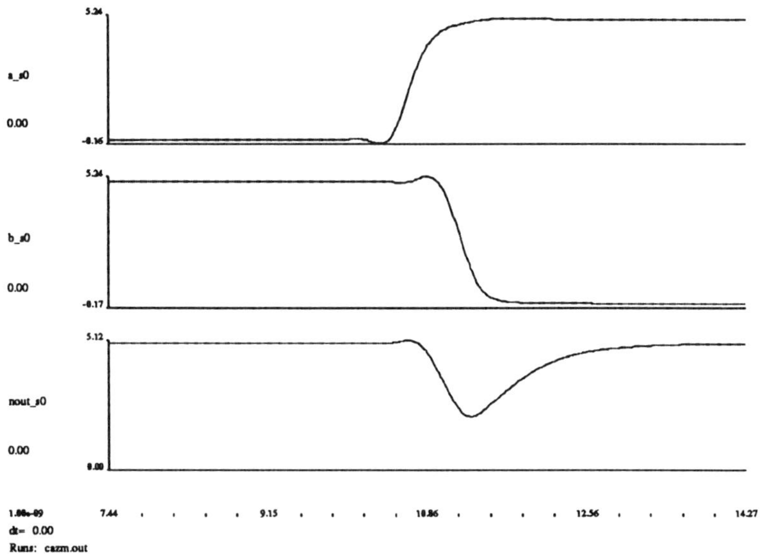

Figure 5.16 Glitch in Output of Two input NAND Gate

glitch is represented as a steady event (assumed nonexistent). Otherwise, the glitch is represented by two transition events (each assumed full swing). The problem occurs when the glitch is propagated to the input of another gate and is combined with an event on the other input terminal. The effect of a partial swing glitch on the delay of the circuit element is not considered.

This type of inherently analog behavior is very difficult to model in a purely digital timing analyzer. In most cases, the best solution is to report these potential glitch cases and verify the timing using circuit simulation. This limitation is, in general, true of any purely digital timing analysis tool.

Given the structure of the *XTV* data representation, however, an extension could be made to better allow for accurate modeling of this case. Currently, there are two types of events that represent the instantaneous state of a circuit node. These are steady events and transition events. Associated with each steady event is a level (high or low) and associated with each transition event is a direction (rising or falling) and an event time. To represent the glitch case, a third event type, the glitch event, could be added. Associated with this event would be an direction (falling/rising or rising/falling), an event time, a height (in volts), and a width (in units of time). Again, this would increase the number of events that must be stored by *XTV* at each element. And, this would also necessitate an extension of the gate models to handle the propagation of glitches. The results, however, would be a much more accurate method of calculating the timing effects of glitches.

5.8.2 Consideration of Variation due to Process/Environment

To this point, the discussion has assumed perfect knowledge of gate delays. In reality, gate delays vary due to process and environmental variation. Typically, gate delays used for timing analysis are worst case delays. That is, when calculating the earliest output event, fastest gate delays are used and, when calculating the lastest output event, slowest gate delays are used.

When analysis is done using graph level algorithms (such as PERT) without considering whether the logic will allow particular events to happen, this type of analysis works very well. A safe bound on the earliest and latest output events can be found. However, when the logic is considered and an attempt to eliminate false responses is made, insuring that the bounds are valid over all ranges of process variation is more difficult. The underlying problem is that responses that were false using worst case fixed delay models at design time can be true with exact delays after fabrication. Thus, for example, when trying to calculate earliest event, it is possible that the earliest event found using fastest delay models may in fact not be the absolute earliest event. That is, slower gate models may yield an earlier event. This phenomena has been identified as the Monotone Speedup Property or the Robustness Property[66]. The reader is referred to [66, 67] for example circuits in which this phenomena is exhibited for simple gate models. In general, this can occur when a path is not dynamically sensitizable using worst case models but becomes dynamically sensitizable when the delay of some circuit element or elements is changed closer to nominal.

When more detailed gate models such as the data dependent models proposed in this work are considered, this robustness property can be important even in circuits without reconvergent fanout and thus no conventional false paths. Consider the circuit shown in Figure 5.17. Assume, as shown in the top diagram, that logic block A produces a single falling transition event at an earliest time of 10 and block B produces a single event at an earliest time of 12. Also, assume that the AND gate shown has a delay of 3 for simultaneous events and 6 for single events. Thus, the output of the AND gate under these conditions is a single falling transition at time 16 (10+6). However, if process variation is considered as shown in the bottom diagram, the output of logic block A may occur any time between 10 and 12 and the output of B may occur any time between 12 and 14. Assuming logic block A produces its latest event while logic block B produces its earliest event, the output of the AND gate could be as early as time 15 (12+3). This is *earlier* than the event time produced using the worst case earliest delays of blocks A and B.

Based on the above examples and as conjectured in [66], it appears that the true earliest and latest event times are inherently nondeterminate properties of logic circuits since precise process information is never known. Thus the designer is faced with an interesting problem when trying to accurately determine the earliest and latest events of a block of logic. There are several strategies for dealing with these concerns. First, the designer may desire bounds on the worst/best case delays in order to insure that timing requirements are met. The least complex solution is to use graph level PERT-type analysis (not considering logic) using true worst and best case gate models. In order to insure that the bound is conservative, simultaneous input transitions must be assumed to be possible. This resulting bound will be safe although it may possibly be too conservative to be useful. Recent research has produced several other algorithms that determine bounds using simple delay models and are also valid for all possible circuits within the process bounds [9, 67, 68, 15, 21, 55]. However, these procedures do not consider the effects of simultaneous input transitions.

To allow for more accurate delay models and the problems demonstrated in the examples above, the *XTV* event representation methodology could be extended to store ranges of delay rather than exact delay values. Then, as shown in Figure 5.17, worst case variation on event times could be propagated through the logic and true bounds on the worst possible delay could be insured. This would only require minor changes to the *XTV* data structures and gate models.

Note that the examples given above assume that the delays of gates in the circuit can vary independently. While this is indeed true with gates fabricated on different chips, it is not as true of gates fabricated on a monolithic device. While, there may be small random local variations or gradient variation across large areas on the same chip, it is unlikely that the delay of two gates in the same section of the chip will vary independently by a large amount. Typically the gate delays will track each other.

Based on this observation, a third strategy is to perform exact timing analysis as outlined in the earlier part of this chapter based on a sampling of timing models from different possible processes. These models would include nominal

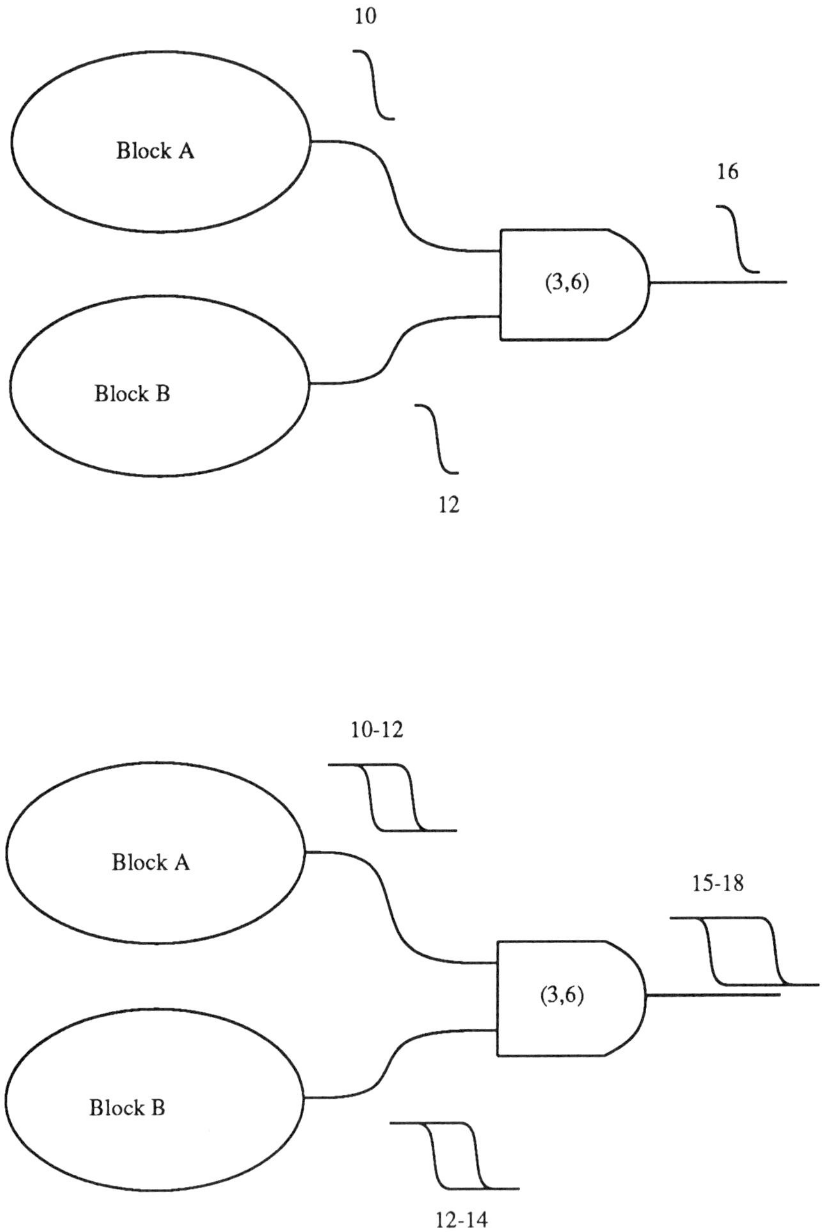

Figure 5.17 The Robustness Property

process timing models, corner process timing models, and a sampling of timing models from the interior of the process space. This strategy can give a desired level of confidence dependent on the number of runs performed. The drawback to this approach, of course, is that many timing analysis runs many need to be performed. However, more realistic bounds on delay will be achieved.

5.9 CONCLUSION

This chapter has presented a new algorithm for exact timing analysis considering data dependent gate delays. This procedure calculates and stores response event times just as the procedure of the last chapter but stores the response constraint implicitly in a response dependency graph. This greatly improves the time needed to do the analysis and thus allows for analysis of larger circuits. This chapter also described limitations of the methodology along with several enhancements and extensions involving the gate models and the ability to predict earliest and latest delays over a range of process variation.

6

PRACTICAL CONSIDERATIONS IN WAVE PIPELINING

The theory and algorithms presented in previous chapters have concentrated on the *analysis* of fixed circuit designs. Theoretical constraints on various timing parameters at the registers and internal circuit nodes have been discussed along with algorithms for the timing analysis of individual combination logic blocks. In both cases, the discussion has concentrated on analysis based on abstract timing and circuit parameters.

The goal of this chapter is to relate the theory of the previous chapters to the physical *design* of practical maximal rate circuits using CMOS technology. This will involve investigating the physical meaning of the abstract timing and circuits parameters used in the theoretical development and the discussion of various design decisions and design techniques for realistic implementation. Although the primary target of this discussion is wave pipelined circuits, many of the techniques described apply to general high performance circuits as well.

From Chapters 2 and 3, it is evident that the most important goal for optimal performance of wave pipelined circuits is the equalization or balancing of maximum and minimum circuit delays (t_{max} and t_{min}). Therefore, the focus of discussion in this chapter is the investigation of various physical sources for this delay variation and how to reduce their effects.

This chapter is organized into six sections. Throughout the discussion, various sources of potential delay imbalance and, thus, potential locations for performance enhancement, will be highlighted. In Section 6.1, the choice of architecture is shown to be of fundamental importance for any wave pipelined application. Next, variation in delays due to differing circuit path lengths is examined. In Section 6.3, various CMOS circuit families are discussed in terms of the amount of inherent gate delay variation. In Section 6.4, variation from process manufacturing is investigated, and practical physical design issues including clock distribution, noise control, and interconnect routing are discussed in Section 6.5. Finally, issues relating to the performance of wave pipelined circuits as technology is scaled are discussed in Section 6.6.

6.1 ARCHITECTURE CHOICE

A major consideration in the design of any system is the choice of overall architecture. As the architecture certainly has a impact on the performance of conventional systems, the architecture can have a major impact on the performance of the wave pipelined circuits. Thus, it is important to investigate the considerations relating to architecture choice in wave pipelined systems and how those considerations differ from conventional systems.

The major consideration from a performance standpoint in the design of conventional systems is the worst case latency of the system since this relates directly to the minimal clock period. With wave pipelined systems, however, the major consideration is the variation in lengths of delay paths. The chosen architecture can have a major impact on the ability and cost of equalizing delays for a given application.

For example, consider the design of a binary adder circuit. The most common and least complicated design is the carry ripple adder. However, the difference between the longest and shortest path is linear ($O(n)$) in the number of addend bits. This implies that, in order to equalize path delays, much padding is needed, and the area penalty for such padding will be $O(n^2)$. As an alternate design of the adder circuit, consider the carry lookahead circuit. The difference between the longest and shortest path in this case is $O(\log(n))$ implying that much less padding will be needed.

There has been much research in "balanced" architectures for design methodologies other that wave pipelining. For example, in standard conventional design, there are many advantages to regular layout architectures. Therefore research has been done to resynthesize architectures for ease and regularity in layout. An example of this work is the adder architecture developed by Brent and Kung [10]. This architecture will be discussed in more detail in Chapter 7. Another research area that has produced regular and balanced layouts is the field of systolic arrays. This architecture style was proposed by Kung and Leiserson[53] and consists of a highly structured, regular array of nearest-neighbor-connected simple processors. While these architectures employ conventional pipelining, many can be modified to support wave pipelining. Systolic array architectures have been proposed for literally hundreds of applications.

Although, the variation in architecture path lengths is obviously the most important criterion for wave pipelining, the latency plays an important secondary role. As with any pipelined system, results are produced on every clock period. However, the total time from initiation of data to the finished result is determined by the latency of the circuit, and, in many applications, this parameter is also important. For, example, if the pipeline must be flushed to begin a new set of computations, the latency of the circuits relates to the amount of startup time before any new results are available. For wave pipelined circuits, latency is simply the delay of the longest critical path. In many cases, the goals of reduced variation and reduced latency may conflict, and a design compromise may be necessary.

Lastly, the existence of feedback in the architecture will produce additional constraints on the clock period as described in Chapter 3. In this case, the total delay around the feedback loops relates directly to the minimum clock period (see Section 3.3.5). This implies that, in general, the maximum delay around a loop should be minimized. Note that the addition of feedback greatly complicates the design of wave pipelined circuits indicating that the easiest circuits to wave pipeline will be the circuits with straight-through data flow.

6.2 PATH DELAY VARIATION

Perhaps the most obvious source of delay imbalance is differences in path lengths. After the architecture is chosen and the high level computational blocks are mapped to a gate level representation, there will, in general, be differences in the lengths of various circuit paths between registers.

In general, this imbalance can be lessened by adding delay elements to the short paths or by adjusting individual gate delays to lengthen short paths or speed up long paths. This balancing has been done by hand in most working wave pipelined designs to date [1, 78, 24, 59]. However, several groups of researchers have developed algorithms for equalizing path delays in combinational circuits by adding active delay buffers [108, 107, 22, 49] and by fine tuning gate drive currents [109, 107].

As described in Chapters 4 and 5 though, path lengths alone are, in general, not enough information to determine true CMOS circuit delays because of the difference in CMOS gate delays due to different data inputs. In fact, as is demonstrated in the next section, path delays can be exactly balanced with still a wide variation in overall circuit delays.

6.3 CIRCUIT CHOICE

The choice of circuit family for use in a wave pipelined system can have a significant impact on the performance the system though the effect on unequal t_{max} and t_{min} at the gate level. This is due to gate delay differences for different input patterns or different signal timing patterns.

This is of particular importance in CMOS circuits and implies that a gate should be chosen that minimizes this effect. Parallel transistor paths, such as in a two-input static CMOS NAND gate (Figure 6.1-A), can give a factor of two difference in circuit delay depending on input data as discussed in Chapter 4 (see Figure 4.2). This effect is even more pronounced in gates with more inputs (see Figure 5.15). Also, serial transistor paths can induce different gate delays. This effect is much less that the rising delay variation from the pullup path but still may be significant for certain layout styles.

Several options are available to reduce the effect of the parallel paths. First, the effect of the parallel paths can be reduced by using a NAND structure proposed in [50] (Figure 6.1-B). By careful transistor sizing, the resistance of the pull-up structure is dominated by series PMOS transistor and thus variation due to the parallel paths is reduced. This structure is claimed to have a rising delay variation of about 20% [50].

Alternately, by using a biased CMOS logic family, the parallel paths can be eliminated altogether by replacing the pull-up structure of the conventional NAND by a single pullup transistor as shown in Figure 6.2-A. This is similar to the pseudo NMOS circuit family in which the gate of the PMOS transistor is permanently connected to 0V. The bias voltage on the PMOS transistor of biased CMOS family gates also allows for some control of the overall delay of the gate and the fine tuning of rising vs. falling transition times. For example, Figure 6.3 shows the rising delay of a biased CMOS inverter vs. bias voltage using minimum sized MOSIS 2μm devices.

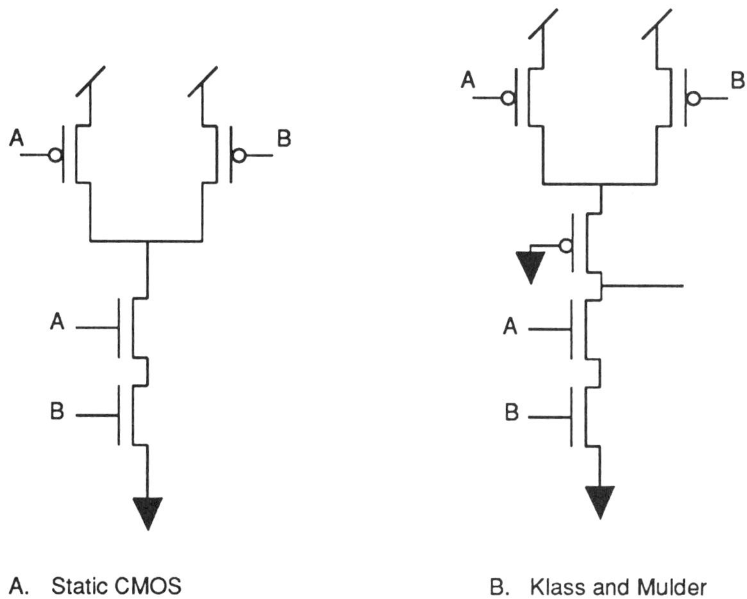

Figure 6.1 Candidate NAND Structures

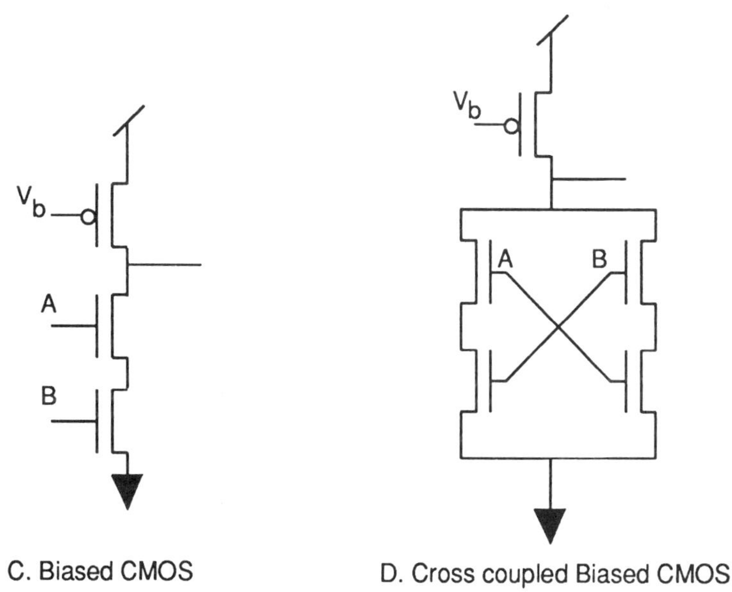

Figure 6.2 More Candidate NAND Structures

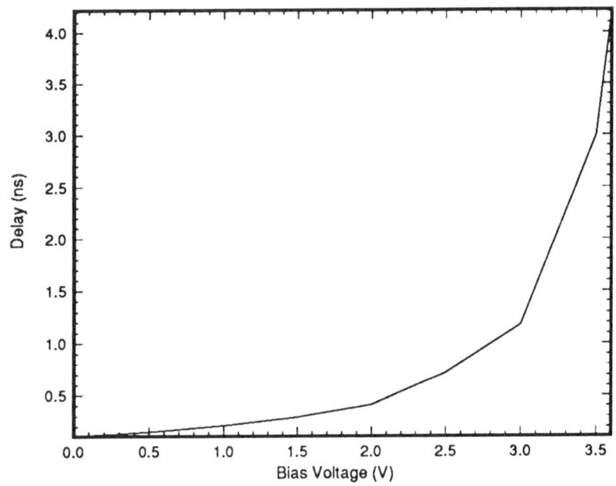

Figure 6.3 Delay of Biased CMOS Inverter

In order to minimize the effect of the serial path imbalance, a cross-coupled biased CMOS NAND (Figure 6.2-B) can be used although at a cost of slightly more area. This cross-coupled NAND structure has been used in the design of a fast wave pipelined adder [24] to be described in Chapter 7.

The design methodology of the biased CMOS logic family can also be extended to differential logic families. For example, Figure 6.4 shows a differential implementation of an exclusive-or (XOR) gate. As shown, the pull-up path consists of the differential active load PMOS transistors and the pull-down path consists of serial paths of NMOS transistors cross-coupled to reduce variation in delay. This differential logic style can be extended to other gate functions as well.

It is important to note that imbalance in delays of the logic gates will fundamentally limit the difference $t_{max} - t_{min}$ even if perfect path balancing is achieved. For example, consider a combination logic block built exclusively of conventional static two-input NAND gates where the minimum delay of the gate (t_{min_g}) is one half the maximum delay (t_{max_g}) for both rising and falling delays of the gate. That is, $t_{min_g} = \frac{t_{max_g}}{2}$. Assuming n logic stages and the existence of worst case patterns, the difference in t_{max} and t_{min} of the circuits would be one half of the delay of the longest critical path thus limiting the

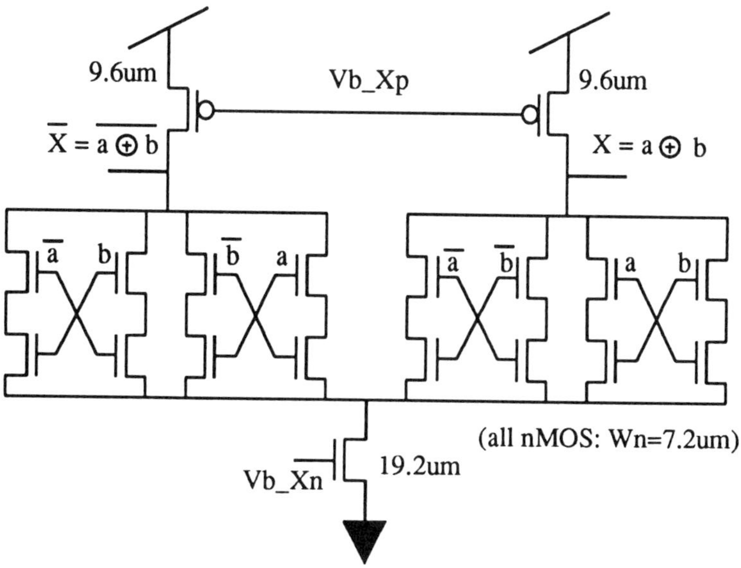

Figure 6.4 Differential XOR

speedup to at most two even if all path lengths are identical.

$$\begin{aligned} t_{max} - t_{min} &= nt_{max\,g} - nt_{min\,g} \\ &= nt_{max\,g} - \frac{n}{2}t_{max\,g} \\ &= \frac{n}{2}t_{max\,g} \\ &= \frac{t_{max}}{2} \end{aligned}$$

This emphasizes the need for the use of gates such as biased CMOS gates with low inherent delay variation and with as few inputs as possible.

A last consideration in the selection of circuit elements arises from the internal node constraints of Section 2.3.2. As shown in constraint 7 of Chapter 2, the minimum clock period is also affected by the parameter $t_{stable}(i)$. As explained, this parameter relates to the amount time that inputs to a gate must be stable in order for the output to fully respond. For simple CMOS gates, this, in turn, relates to the gate drive and the output loading of the gate implying that excessively slow gates or heavily loaded nodes due to long interconnect or high fanout should be avoided. Thus, combinational logic should be implemented using multiple level logic with smaller (faster) gates rather than in two-level logic using larger (slower) gates.

6.4 PARAMETRIC VARIATIONS DUE TO MANUFACTURING AND ENVIRONMENTAL FACTORS

In addition to the variation in circuit delays cause by design due to architecture, path delay differences, or data dependent circuit delays, differences in t_{max} and t_{min} are caused by the integrated circuit manufacturing process and by variation in the ambient environment while the circuit is operating.

6.4.1 Sources of Parametric Variation

Unfortunately, there are many sources for these delay variations that must be considered especially considering the high operating frequencies that the wave pipelining technique makes possible. These sources can be categorized into two major groups: ***static variation*** (for example, due to fabrication process) and ***dynamic operational variation*** (for example, due to temperature changes).

The static component refers to the inevitable process variations that occur between chips of different fabrication runs, between chips at different locations on the same wafer, and between transistors at different locations on the chip. Slight variations on channel lengths of fabricated devices can produce variations in nominal delay of the logic gates which could cause phase misalignment at a latch. If the delay units are made programmable, this static component can be compensated for after fabrication although at a cost of more expense testing and higher system cost due to the need to calibrate individual chips.

The dynamic component is caused by the variation in gate delay induced by changes in the operating environment of the device. For example, the amount of power will vary widely depending on whether the data causes many devices to switch or few devices to switch. This can, in turn, cause variations in temperature and thus the delay of the device. Other sources include random noise, jitter of clocks and data, and variations in power supply voltage.

As with the static variation, this dynamic variation must be considered in t_{min} and t_{max} or compensated for after fabrication using structures such as phase locked loops and dynamically programmable delays. This type of compensation has been considered in the design of a high speed wave pipelined FIFO[29].

6.4.2 Toleration of Parametric Variations

Since many of the sources of both static and dynamic parametric variations are not easily quantifyable before fabrication, it is interesting to determine how much variation can be tolerated by the wave pipelined circuit if these variations are not included in t_{max} and t_{min}.

From Figure 2.9, some interesting observations can be made concerning operation in different k regions. Consider an operational frequency (T_{clk}) between $t_{max}/2$ and t_{min}. Depending on the length of the delay unit (Δ) the circuit can operate with $k = 1$ or $k = 2$. Note, however that there is a major difference between the two cases in terms of the amount of variation in clock frequency that can be tolerated. With $k = 2$ and $\Delta = 0$ the frequency must be between $t_{max}/2$ and t_{min} for correct operation. However, in the $k = 1$ case with $\Delta = t_{min}$, the frequency range is only bounded below by $t_{max} - t_{min}$. This implies that systems operating with lower k will be able to tolerate more variation in frequency and will thus have a greater range of operation. This, however, is at the expense of larger delay units.

This same effect can also be observed when process and environmental variations on combinational logic path delays are considered. Assuming a constant scaling factor c (per chip) representing the amount of delay variation from nominal values due to process and environmental effect, the delay parameters t_{min}, t_{max}, and Δ become ct_{min}, ct_{max}, and $c\Delta$. This will modify the feasible constraint space of Figure 2.9. An upper bound (assuming $t_{max} = t_{min}$ and $\Delta = 0$) on the maximal amount of variation, c, that can be tolerated for a given k such that correct operation is maintained is $\frac{1}{2k-1}$ (for $k > 1$). This relationship is derived in Appendix B.

Thus, if $k = 10$, under ideal conditions only approximately 5% variation in c can be tolerated. This implies that wave pipelined operation in high k modes will be very difficult. Note, however, that, by appropriate choice of Δ, the circuit can be operated in a lower k region without increasing T_{clk} thus allowing for maximum process/environmental variation. Notice also that in the $k = 0$ and $k = 1$ regions a direct tradeoff can be made between the amount of variation that can be tolerated and the T_{clk} of the system since the feasible clock period is unbounded above in these regions.

The considerations outlined in this section imply that, for reliable wave pipelined operation, lower k should be used along with careful consideration of environmental and process variations.

6.5 CLOCK DISTRIBUTION AND PHYSICAL LAYOUT

The design of the clock distribution network for high speed wave pipelined circuits is critical. It is easily seen from constraints (4) of Chapter 2 and (4) of Chapter 3 that any unintentional clock skew at any register (either early or late) will cause a direct increase in the minimal clock period. This is captured by parameters Δt_e^i and Δt_l^i for each register i. Thus, it is important that precise routing and noise control be employed to minimize this skew.

Also, many times it may be desirable to add intentional delay along the clock path to optimize clock skews as described in Chapters 2 and 3. This requires the insertion of precise delay elements. In light of the discussion of process variation of the previous section, it may be desirable that these inserted delay elements be controllable either initially after fabrication or even during operation. These concerns are current research topics in their own rights and are also beginning to be considered even in the design of conventional systems. For example, actively controlled phase lock loop circuits are being employed in current microprocessors to eliminate clock skew and produce zero-skew clocking networks [37, 114, 33]. This type of approach can also be used in wave pipelined systems to produce *non-zero* constant skew.

A second concern relating to physical layout is the routing of signal interconnect. The parasitic capacitance and resistance of interconnect affect the timing of wave pipelined systems in two ways. First, interconnect capacitance affects the loading of logic gates and thus their relative delays. Since the goal is to balance delays to the outputs, interconnect must be routed such that the correct amount of loading is maintained relative to other logic gates on other paths. Second, depending on the relative values for the resistance and capacitance of a long interconnect signal the RC delay of the routed line may become significant. In this case, it may be necessary to balance both capacitance and resistance of signal lines. Lastly, the loading of the gate affects the $t_{stable}(i)$ of circuit nodes.

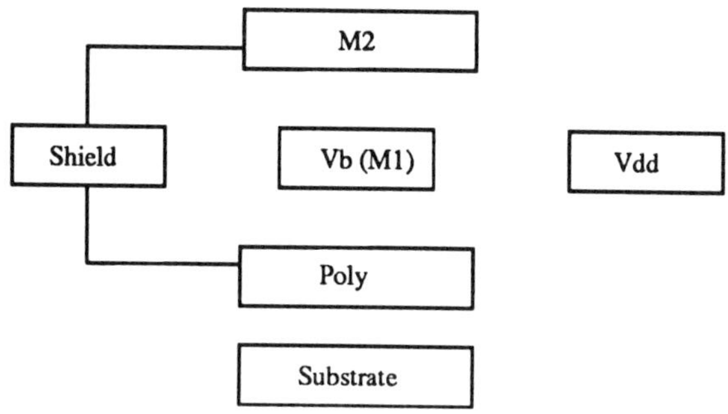

Figure 6.5 A conceptual cross-section of the cross talk prevention structure

Thus, an obvious goal is the minimization of interconnect loading or the insertion of larger drivers for large interconnect. Again many of these requirements, especially the requirement to minimize loading, are significant in the design of any high speed system.

In the previous section, the concern for the effects of dynamic parametric variations in the design of wave pipelined circuits was raised. Many times these concerns are referred to under the general topics of noise or jitter. While these topics have traditional been the concern of analog designers[31], high speed digital design, including wave pipelined circuits, are becoming increasingly affected by these phenomena. Many of the structures and design techniques used by analog designers can be used in digital system to reduce noise and control jitter. These include shielding of interconnect, physical isolation between sensitive circuitry and potential noise sources, and the use of circuit techniques such as differential circuits that produce less switching noise and reject more power supply noise. An example of the type of noise prevention structure that has been used in the design of the parallel adder (described in the Chapter 7) is shown in Figure 6.5. This structure was used to prevent noise from coupling to a reference bias voltage used to control the delay of biased CMOS circuits as shown in Figure 6.2-D. As shown the metal 1 bias line is shielded above and below by "quiet" shield lines in metal 2 and poly and on either side by the shield line and a Vdd line.

6.6 TECHNOLOGY SCALING

Since the performance of circuits designed using the standard static CMOS design methodology steadily increases as feature sizes are scaled down, it is natural to ask how the performance of wave pipelined systems is affected as feature sizes shrink. On the surface, it seems simple to assume that, if absolute gate delays scale down by some factor, s, then the difference in sums of delays would also scale by the same factor. However, the difference in absolute gate delays is only one factor in the overall t_{max} and t_{min} of wave pipelined circuits. Other factors such as random process variation, dynamic jitter, and data dependent gate variation do not necessarily scale at the same rate as the absolute gate delay. Thus, overall performance of wave pipelined circuits may not increase at the rate of conventional circuits as feature sizes are reduced.

Still, scaling does have a positive effect on many components of the wave pipelining performance limits. First, scaling will reduce the difference in delays when there are significant differences due to different circuit path lengths. Second, data dependent delay variation may decrease dependent on the actual process technology. As mentioned in Section 4.2.2 of Chapter 4, the difference in delay for a particular gate typically depend on parasitic capacitances that are typically reduced as the technology scales. It is important to note that the rate of scaling depends on the exact technology process. And lastly, technology scaling will increase the performance of conventional components that are used in conjunction with the wave pipelined circuit. For example, the speed of registers or latches used to clock data into the circuit can be increased. Also, chip input and output buffers could be made faster. Thus, in general, wave pipelined design will benefit from technology advancement.

6.7 CONCLUSION

This chapter has investigated several practical considerations in the design of wave pipelined circuits. It has been shown that careful design starting with the selection of the architecture is crucial to success. In particular, the choice of base level circuit family can significantly limit the performance of wave pipelined systems even if all other design choices are flawless. These considerations suggest that careful design is necessary but also that wave pipelined design can be realized.

Many of the concerns described have traditionally been the exclusive concern of analog designers. However, it is becoming increasingly evident that they must be considered for any successful high speed design. In particular, for high speed wave pipelined design, concerns such as process variation, noise, and jitter must be addressed from the early design phases to insure success.

7

DESIGN EXAMPLES

7.1 INTRODUCTION

The wave pipelining methodology represents a significant departure from conventional circuit design and conventional pipelining by deriving speed from balancing or matching delays in combinational logic rather than simply minimizing delays. Therefore, it is important to show the feasibility of this technique through the fabrication of actual devices. As previously mentioned, other research has involved fabricating wave pipelined circuits however most of this research has been in technologies other than CMOS or has not pushed performance to its theoretical limit.

This chapter presents the design, implementation, and test of two high speed CMOS integrated circuit designs using the concepts of wave pipelining. The first design is a 16-bit parallel-carry adder implemented in MOSIS 2μm CMOS process. This design is fully wave pipelined, operating at a maximum rate of 250 million additions per second with up to nine virtual pipeline stages. The second design is a digital sampling circuit capable of 1 Gbit/s bandwidth and 25 ps sampling resolution. This design is not wave pipelined by the classic definition of wave pipelining but does rely on the propagation of multiple signal "waves" through the data and clock paths and demonstrates several important characteristics common to wave pipelining. This design was implemented in MOSIS 1.2μm CMOS process. Both designs push the concepts of wave pipelining to the theoretical limit.

Throughout the discussion of the chapter, design decisions and results of these designs will be related to the concepts presented in previous chapters. For both test chips, the discussion is organized into three subsections. First, fundamental design concepts of the application are presented. Next, the implementation of the design is described including detailed circuit and physical design. Finally, test results from the fabricated device are presented and are related to theoretical wave pipelining concepts. For additional information on details of the design and test results of these integrated circuits, the reader is referred to corresponding technical reports and papers [24, 61, 100, 102, 35].

7.2 PARALLEL-CARRY ADDER

In this section, the design, implementation, and test of a 16-bit parallel-carry adder is presented. The adder is a fully wave pipelined single-stage system of the type described in Chapter 2 with no registers internal to the combinational logic block. At its maximum operating speed, the adder can support nine virtual pipeline stages.

7.2.1 Design

As mentioned in Section 6.1, the circuit architecture can have a major impact on the performance of a wave pipelined circuit. In that section, an example was discussed comparing a carry-ripple adder to a carry-lookahead adder. It was shown that the carry-lookahead adder is much nearer balanced and is thus more suitable for wave pipelined operation since optimal wave pipelined performance depends on balanced circuit delays.

The adder structure implemented in this test chip contains a full carry-lookahead structure and was derived from a regular-layout parallel adder proposed by Brent and Kung[10]. In order to save overall area and power, the Brent/Kung architecture was modified such that the logic depth was reduce to one half of the original at an expense of larger drivers because of greater fanout. The new adder architecture and its carry computation are shown in Figures 7.1 and 7.2 respectively.

For the proposed tree structure, a set of recurrence relationships are defined over addend bits, a_i and b_i, for every i, as follows:

$$g_i = a_i \wedge b_i \qquad (1)$$

$$p_i = a_i \oplus b_i \qquad (2)$$

$$c_i = G_i, \quad for\ i = 1, 2, \cdots, n \qquad (3)$$

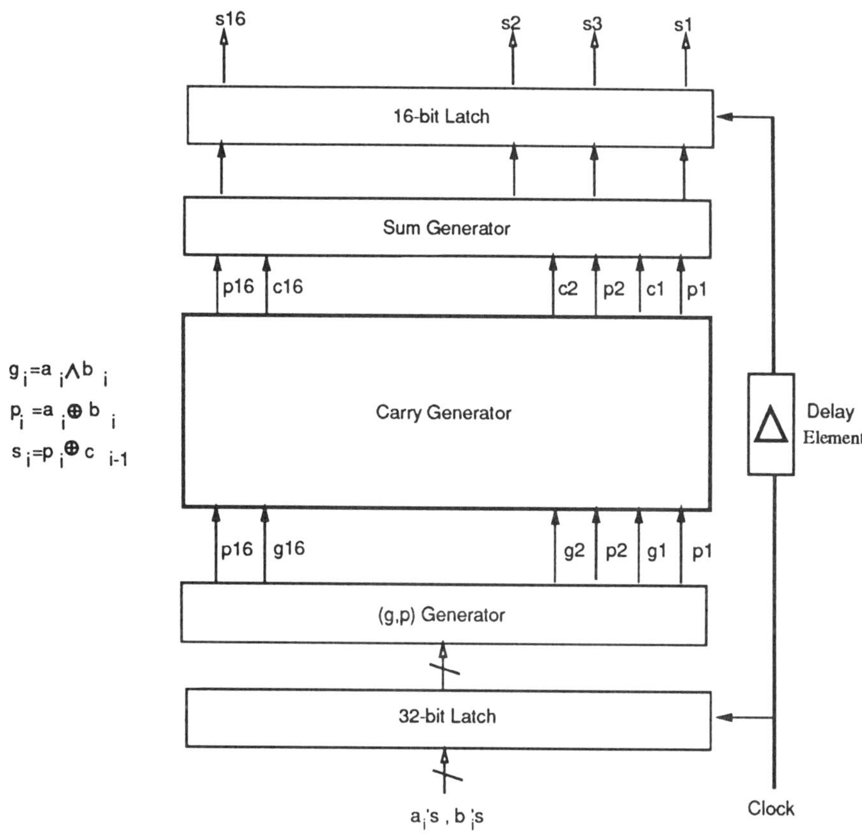

Figure 7.1 Block Diagram of Parallel Adder

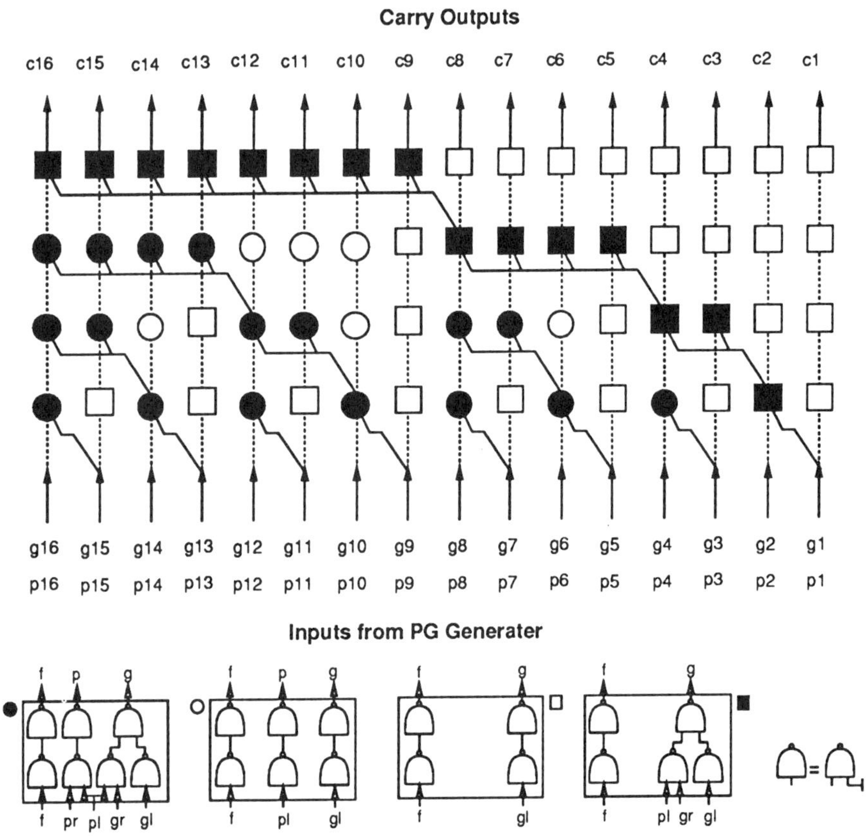

Figure 7.2 Computation of all the carries for n = 16

where

$$(G_i, P_i) = (g_1, p_1) \quad for\ i\ =\ 1 \tag{4}$$

$$(G_i, P_i) = (g_i, p_i) \circ (G_{i-1}, P_{i-1}) \quad for\ i\ >\ 1 \tag{5}$$

and $\circ$ is a concatenation operator defined as follows:

$$(g_l, p_l) \circ (g_r, p_r) = (g_l \vee p_l \wedge g_r, p_l \wedge p_r) \tag{6}$$

After the completion of carry bits c_i, the sum bit s_i are computed by

$$s_i \quad = \quad p_i \oplus c_{i-1} \tag{7}$$

$$s_1 \quad = \quad p_1 \tag{8}$$

Three circuit blocks, shown in Figure 7.1, are required to implement the adder functions defined above. The first circuit (labeled (g, p) Generator) takes adder inputs and generates the p_is and g_is as in Equations (1) and (2). Then the p_is and g_is are fed into the Carry Generator for the computations of c_is (Equation (3)). The third block, the Sum Generator, consists of a row of XOR gates which depend on the c_is and p_is as shown in Equation (7).

As shown in Figure 7.2, the carry structure is implemented with four basic modules: black circle, black square, white circle, and white square. The black modules perform the useful computation defined by Equation (5), and the white modules simply serve as "padding elements" for passing inputs to outputs with appropriate delays. In this way, the delays are balanced at each logic level and the data associated with each wave propagate coherently through the level logic. Similarly, all p_is propagate coherently through f paths in the black and white modules so that they arrive concurrent with corresponding carry bits, c_is, at the inputs to the Sum Generator. The circle module shown in Figure 7.2 produces p, g, and f, while the square module produces g and f.

7.2.2 Implementation

Circuit implementation and physical design are of paramount importance to the performance of any high speed digital system including wave pipelined designs. In the implementation of the adder circuit, four major areas were considered. These were circuit level implementation of logic gates, minimizing the effects of process and environmental variation, physical layout, and apparatus for testing after fabrication.

Circuit Design

The ideal gate for wave pipelining should have no delay variation due to different input patterns thereby not contributing to the difference in maximum and minimum delays through the logic. Although CMOS technology is very attractive for this application in terms of cost and integration capabilities, conventional static CMOS has severe data dependent delay variation as has been mentioned in Chapters 4 and 6 due to parallel transistor paths and, to a lesser extent, serial transistor paths. In addition, static CMOS has high input capacitance, large feedthrough current, and high dynamic temperature variation.

To overcome these problems, the adder circuit was implemented entirely of two-input cross-coupled biased CMOS NAND gates as shown in Figure 7.3b. When inverters are needed for computation or padding, NAND gates with one input tied to Vdd (hence inverters) are used. This approach allows the adder implementation to use only one type of gate (NAND) and greatly simplifies the task of delay balancing.

As described in Section 6.3, the pull-up paths of conventional static CMOS have been replaced by a single biased PMOS transistor to eliminate variation in delay associated with the standard CMOS parallel paths, and the pull-down path is cross-coupled to reduce the variation due to the serial path. There is a slight area penalty for using the new NAND gate since the number of transistors in a new NAND gate has increased from four to five. However, the data-dependent delay variation has been substantially reduced. Inverters of the type shown in Figure 7.3a were used as buffers between logic levels to drive fanout and interconnect loads.

Balancing Methodology and Physical Layout

Using the biased CMOS NAND gates, the number of gates in every path from input to output was made identical. However, there were differences in fanout and interconnect capacitance that would have still caused variations in delay. To cope will this concern, detailed balancing was done at the layout level.

The goal of balancing at the physical layout level was to reduce delay differences due to gate loading variation, process variation, and environmental variation. For the adder, a conservative design approach was used to have maximal process tracking and thus minimal difference in path delays. Interconnect is routed such that identical loading capacitance (contributed by routing and gate fanout capacitance) is achieved at each bit position in the same stage of the adder.

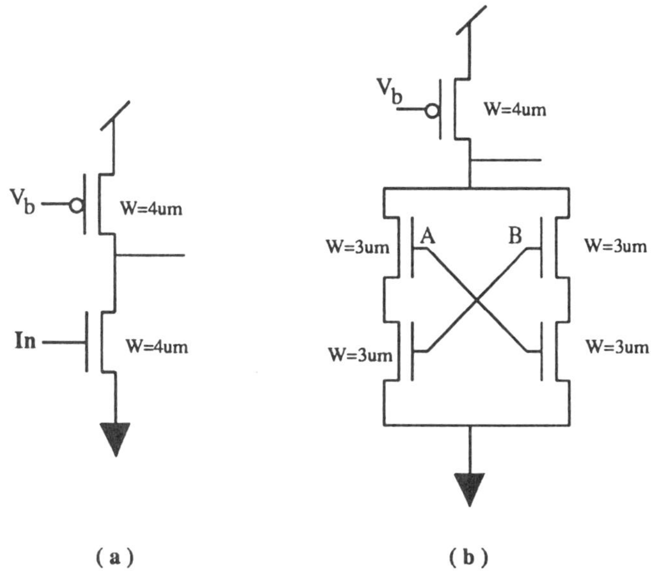

Figure 7.3 Components for adder circuit (all transistor lengths are 2 μm): (a) biased inverter (b) biased NAND gate

Table 7.1 Buffer Sizes (μm) and Approximate Capacitive Loading

Level	Buffer 1 (w/l)†	Buffer 2 (w/l)†	Buffer 3 (w/l)†	Buffer 4 (w/l)†	Capacitance (fF)
pg - 1	3/2	6/2	12/2	25/2	86
1 - 2	3/2	6/2	12/2	24/2	140
2 - 3	4/2	8/2	16/2	40/2	270
3 - 4	4/2	10/2	24/2	60/2	535
4 - sum	3/2	6/2	12/2	24/2	74

† PMOS and NMOS transistor sizes are identical.

Consequently, identical buffers are deployed at each bit position. In order to adequately drive the load at each stage, these buffers consist of four incrementally sized biased CMOS inverters of the type shown in Figure 7.3.a. Buffer sizes are shown in Table 7.1.

In order to achieve maximal tracking of loading capacitance over process variation, capacitance is matched categorically (via versus via, poly versus poly, first metal versus first metal, gate versus gate, etc.) in the routing process. That is, each interconnect at a particular stage of the adder has the same length and width of polysilicon, first metal, second metal, and the same number of vias. Also additional coupling capacitance is added such that each interconnect has the same amount of overlap, sidewall, and fringe capacitance. This includes the capacitances of first metal to second metal, first metal to itself, first metal to poly, metal to substrate, etc. In order to cope with variations in transistor gate loading capacitance due to differences in fanout at the outputs of a particular stage of the adder, fanout capacitance made of transistors with source and drain tied to ground is attached to all paths other than the one with the maximal fanout.

The measures described above were done manually and have a significant area penalty of approximate 40 percent. Nevertheless, this layout strategy provides the maximal tolerance of process variation in the loading capacitance. It should be noted that this strategy was adopted because of relatively loose process specifications provided with the MOSIS 2μm CMOS process. With better information on process parameter distributions, better process control, and more accurate parasitic extraction, it may be possible to use a less rigid matching approach. For example, instead of maintaining equal routing length, various sizes of transistors could be deployed to drive various lengths of routing capacitance. This would substantially reduce the area penalty and power usage.

To reduce the effect of dynamic environmental variation on the delay, measures were taken to reduce cross talk, minimize induced noise on the bias wire (Vb), and reduce power supply noise. Cross talk between signal lines is minimized by avoiding routing signal lines in parallel in close proximity as much as possible and by minimizing the number of crossings of a signal line over another signal line. Where two signals must be routed in parallel for a long distance, a ground wire is placed between the two signals. In order to minimize induced noise on the bias wire (Vb), shield lines are placed above, below, and beside the bias line as was described in Chapter 6 (Figure 6.5). Dynamic fluctuation in power supply levels is reduced by inserting on-chip decoupling capacitors between power and ground planes. Also, many power/ground pins are deployed to further reduce the ground bounce noise.

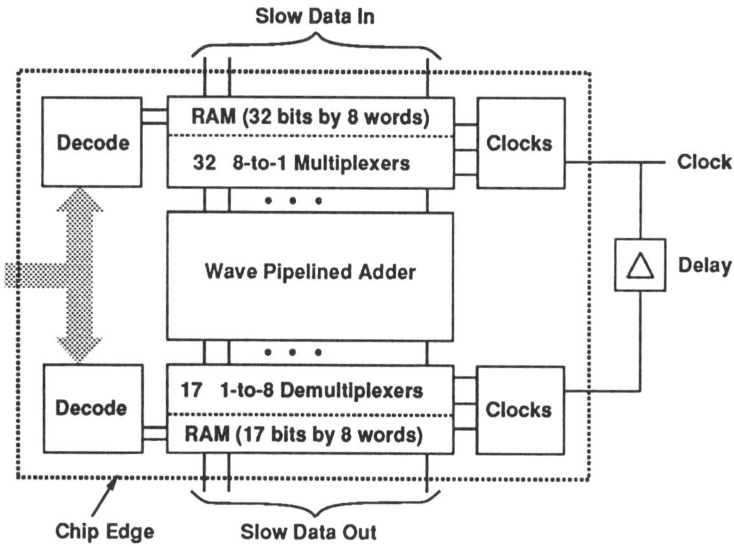

Figure 7.4 Block Diagram of the Chip

Test Structure

The last major area of concern in the implementation of the adder was the ability to test the adder circuit at the anticipated speeds of over 200 MHz. Since these I/O speeds are difficult in 2μm CMOS and the availability of high speed test equipment was limited, a test structure was implemented on-chip. This test structure consists of an 8:1 multiplexer for each input bit and 1:8 demultiplexer for each output of the adder. The input of the multiplexer and the output of the demultiplexer are RAMs that allow the data I/O of the chip to be loaded and unloaded at slow speed. Thus, the chip only requires two high speed clock inputs for input and output registers and no high speed outputs. In operational mode, the eight RAM cells for each input bit are read in sequence and multiplexed to the adder inputs. At the outputs, results from each bit are demultiplexed and stored in successive RAM locations. A conceptual block diagram of the entire chip including the test structure is shown in Figure 7.4.

The clock generation for the multiplexers and demultiplexers is accomplished with a tree of toggle flip flops. The main cell takes a single phase clock and produces a delayed version of itself along with two half frequency clocks lined up with the edges of the delayed clock. One clock at half frequency aligns with the

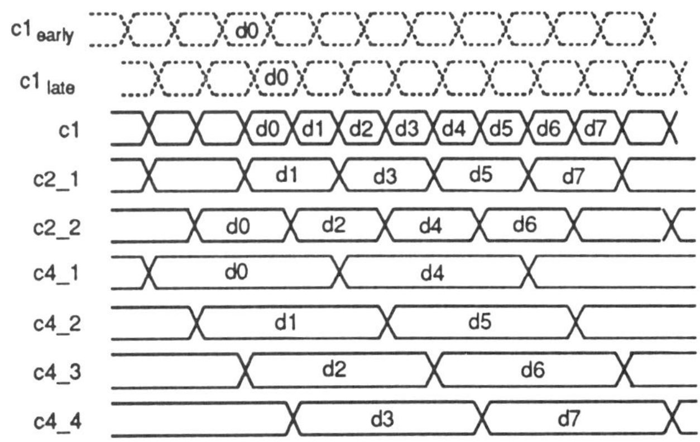

Figure 7.5 Timing of the Multiplexer Circuits: Timing Diagram of Clocks

positive edge of the delayed clock and the other half frequency clock aligns with the negative edge. This is accomplished with matched delays. The clocks, c1, c2_1, c2_2, c4_1, c4_2, c4_3, and c4_4 of Figure 7.5 are generated by the circuits shown in Figure 7.6. This figure also shows the relation of these clocks to the high speed serial data, d0 $\cdots$ d7, generated by the multiplexer. Note how d0 is passed to the output. From this timing diagram, it is clear that there is ample skew tolerance between c2_2 and c4_1 in both directions. However, there is no skew tolerance between c2_2 and c1 in the c1-late direction. A programmable delay line is used to compensate for the lack of skew tolerance in the highest speed clock. Figure 7.6 shows how c1 is affected by the programmable delay line.

Two clock inputs are provided to the chip so that the skew between the adder inputs and output can be set off chip. Differential inputs are provided for the two clocks. Duty cycle can be controlled by adjusting the reference voltage input of the differential amplifier or by adjusting the edge placement of the clock generated off chip.

The multiplexer, shown in Figure 7.7, accepts eight parallel bits from a RAM and time multiplexes them into a serial data stream. One bit is output for each transition of the clock thus achieving a bit rate twice the clock frequency. That is, a 100 MHz clock controls 200 Mbit/s data. The circuit operates by steering data through a tree structure of 2 to 1 multiplexers. The control consists of the seven clocks generated on chip as described above.

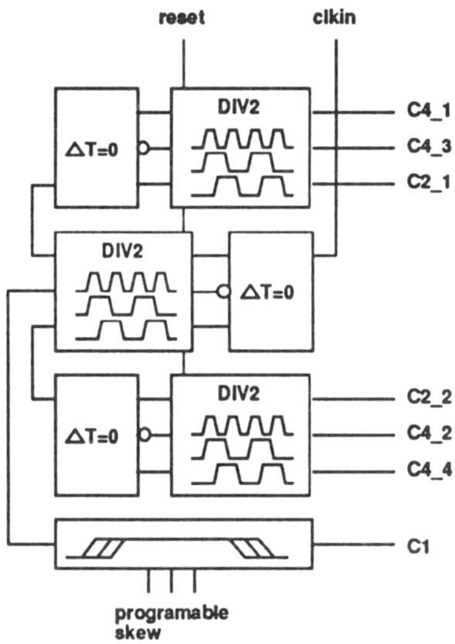

Figure 7.6 Timing of the Multiplexer Circuits: Clock Generation

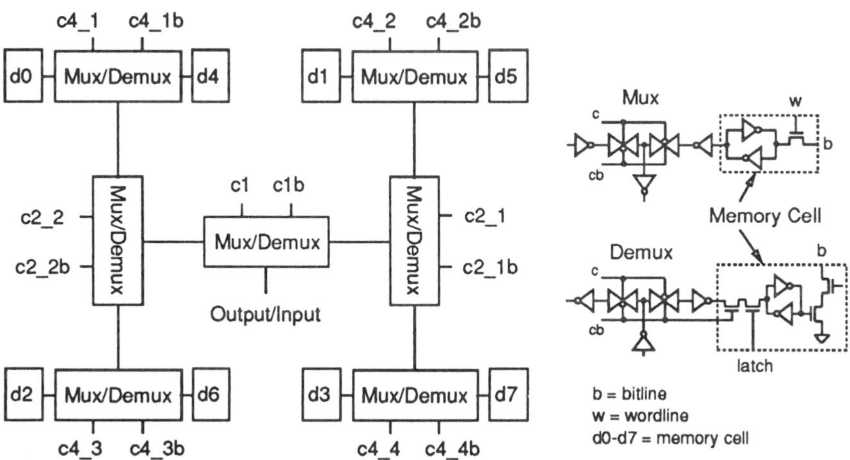

Figure 7.7 Multiplexer/Demultiplexer Block Diagram

Table 7.2 Design Statistics: Adder

Transistors (adder)	6067
Transistors (adder & tester)	15056
Adder Size	3508μm$\times$3782μm
Die Size	4600μm$\times$6800μm
Total IO Pins	106
Power/Ground Pins (adder)	16
Power/Ground Pins (tester)	27
Process	MOSIS 2μm PWELL
Package	108 pin PGA
Power	3 W

The demultiplexer circuit (Figure 7.7) is similar to the multiplexer with the data direction reversed. A less obvious difference is the RAM cell. The eight outputs of the demultiplexer are skewed in time by the pulse width of c1 (PW_{c1}). This requires that the simple RAM cell of the multiplexer be replaced by a clocked register that allows only latched data to be written into the cell. Since the output of the multiplexer is periodic with period $8 \cdot PW_{c1}$, every eighth cycle the same data gets written to a given word of the demultiplexer's RAM. A global line labeled latch in Figure 7.7 allows data acquisition to be started or stopped at any time. The pull-down output of the RAM avoids the possibility of flipping the data during a read. The same clock generation circuitry is used for controlling the demultiplexer.

7.2.3 Test Results

The adder chip was fabricated using the MOSIS 2μm PWELL process. Some design statistics are given in Table 7.2 and a photomicrograph of the actual fabricated device is shown in Figure 7.8. The MAGIC layout system[65] was used for physical design and the CAzM simulator[23] was used for SPICE-type circuit simulation. Of twenty chips received from MOSIS, 17 had no detectable faults at 10 MHz while 16 worked with various degrees of reliability at 200 MHz. Several chips worked up to 250 MHz but with reduced tolerance to power-supply and bias fluctuations. Testing was performed using a HP82000 chip tester at the Microelectronics Center of North Carolina (MCNC).

Figure 7.8 Photomicrograph of Parallel Adder Chip

According to the theoretical results of Chapter 2, two main sets of constraints limit the clock frequency and insure correct clocking for single stage wave pipelined systems. These are the register constraints and the internal node constraints.

As mentioned in Section 2.3.2, the internal node constraints can be assumed to be satisfied under certain conditions. The first condition (condition (*i*) of Theorem 2.1) was that the needed stable time ($t_{stable}(i)$) at the inputs to each gate must be less than the setup and hold time of the registers. Since much effort was made to insure that all transitions at all circuit nodes were similar by inserting incrementally increasing buffer sizes to drive any large interconnect, this requirement was satisfied at design time. The second condition (condition (*ii*) of Theorem 2.1) was that the difference between maximum and minimum delays at each node in the circuit must be less that the difference at later points in the circuit. This fact was also verified at design time. While these conditions cannot, in general, be guaranteed in the fabricated device due to uncertainties in process variation, test results show good correlation to expected theoretical results indicating that the internal node constraints were, indeed, not the limiting constraints.

To determine the register constraints, a two-dimensional constraint space of clock skew and clock frequency was produced as discussed in Chapter 2 and shown in Figure 2.9. This was obtained by performing a pass/fail test on the chip for all vectors in the vector test set while varying clock skew and clock frequency. Near linear behavior is exhibited as the theory predicts. Figure 7.9 shows a measured constraint space for one of the fabricated chips. Note that t_{max}= 38.8ns and t_{min}= 35.5ns for this chip.

The choice of test vectors for the adder was important since the variation in delay due to these test vectors will indicate a large amount of the variation in t_{max} and t_{min}. The test vector set consisted of 800 additions in a fixed sequence. One-hundred and fifty-eight vectors were generated by MikeII[2] to guarantee 100% transition fault coverage. Six vectors were manually created to produce the expected earliest and latest transitions due to data dependent delays. The remaining 636 vectors were randomly generated. These vectors produced delay differences that were consistent with expected results. However, it should be noted that guaranteeing worst/best case coverage across process variation is still somewhat of an open problem as described in Chapter 5.

Although the *XTV* timing analysis program (described in Chapter 5) was not available during the test of the adder chip, worst case vectors produced by *XTV* have since been compared to manually produced worst case vectors using

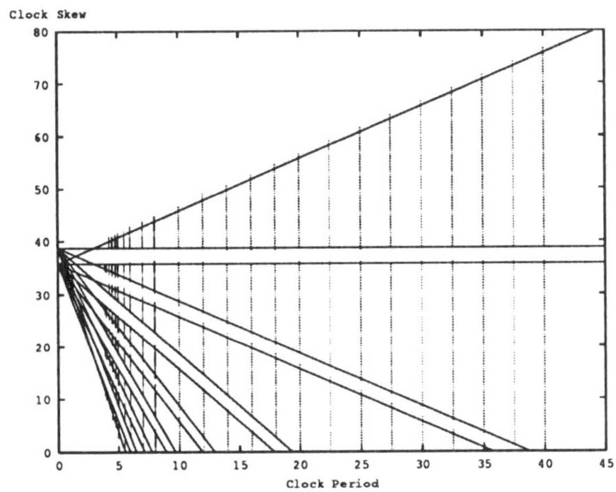

Figure 7.9 Timing Constraint Space of Adder

Table 7.3 Comparison of Manually Created Worst Case Vectors to *XTV*

	Manual (ps)	*XTV* (ps)
Earliest Rising	†	7920
Earliest Falling	8020	7920
Latest Rising	8570	8740
Latest Falling	†	8740

† Not generated manually.

prefabrication timing models (not necessarily calibrated to the fabricated chip). These results are shown in Table 7.3. As shown, manually created vectors are with 2% of the true worst case vectors reported by *XTV* for these timing models. This gives an indication that vectors in the test vector set yielded near worst case results. Note that the actual delay values are smaller than the test result values since the analyzed circuit did not contain the four-inverter buffers between stages.

In order to determine more precisely the sources of the delay variations, additional extensive tests were done to determined the delay variation due to process variation and design imbalance, dynamic clock jitter, variations in bias

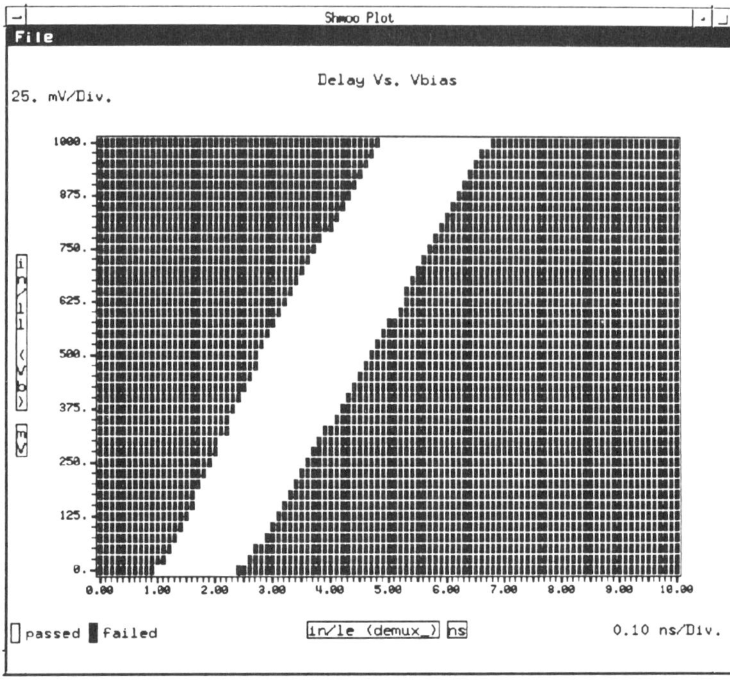

Figure 7.10 Shmoo Plot 1

voltage, variations in power supply levels, data dependent gate delays, and the on-chip test structure. These results are summarized below.

$\mathbf{t_{valid}}$ of Adder

To measure the difference in t_{max} and t_{min}, the HP82000 was configured to measure the valid time ($\mathbf{t_{valid}}$) under different conditions by performing pass/fail tests using the test vector set while varying the output data clock skew with respect to the input data clock. Figures 7.10 and 7.11 show the typical output shmoo plots *delay vs. Vb* and *delay vs. Vdd* respectively for a data rate of 200 MHz. (The white squares corresponds to *PASS*.) Measuring $\mathbf{t_{valid}}$ gives the same information as measuring t_{max}- t_{min} since $\mathbf{t_{valid}} = T_{clk}$-($t_{max}$-$t_{min}$) and T_{clk} is known.

With Vdd = 5V, the maximum $\mathbf{t_{valid}}$ over all Vb is recorded as M1 and S in Table 7.4. M1 includes the timing variations of the test structure. C4 removes

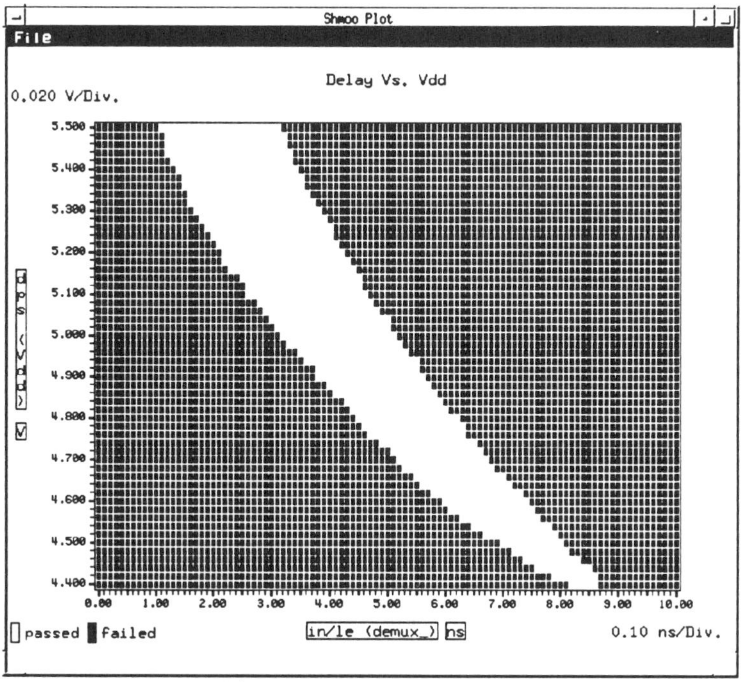

Figure 7.11 Shmoo Plot 2

the effect of the test structure, while C5 estimates data dependent delay by subtracting the following measurement of process variation.

Process Variation and Design Imbalance

To examine process variation, the delay of the sixteen straight-through paths of the adder were determined by stimulating the inputs with single events to eliminate the effect of data dependent delays. The variation across these paths on a chip by chip basis ranged from 0.2 ns to 1.4 ns. There was no correlation between the delay profiles across chips. This indicates that most, if not all, of this variation is due to process. This result is tabulated as M4 in Table 7.4.

Jitter

To find out how much jitter the clock and data exhibited, tests were run on individual paths stimulated by single events to eliminate data dependence and process variation. Measuring the same path delay repeatedly, the degree timing consistency could be observed. With 100 repeated measurements, the delays fell just within a 200 ps window. This result indicates jitter from all sources is about 200 ps.

t_{valid} of Test Structure

The test structure contributed a significant portion of t_{max}- t_{min} of the system. Setup and hold time, clock duty cycle imbalance, and clock jitter contributed to the timing variations in the on-chip test structure.

The valid time of the on-chip test structure bypassing the adder was measured at 200 and 250 MHz. Table 7.4 shows these results. The performance of the test structure quickly degraded near 250 MHz. Hence, testing the adder above this speed was difficult and unreliable.

Clock Duty Cycle Imbalance

Since the latches in the multiplexer and demultiplexer use both rising and falling edges of the clock cycle, the duty cycle of the clock will affect performance of the test structure. By running a three-dimensional shmoo test varying the duty cycle of the multiplexer and demultiplexer clocks while skewing the demultiplexer clock with respect to the multiplexer clock, several of the 20 chips were tested to see the effect of clock duty cycle imbalance on t_{valid} . Figure 7.12 shows a typical surface plot computed from this measurement. The X and Y axes are the pulse widths of the first "half" cycle of the multiplexer and demultiplexer clocks. The Z axis is the valid time of the test structure bypassing the adder. This plot shows that the optimal skew is 300 ps away from the nominal duty cycle of 50%-50%.

Multiplexer and Demultiplexer Uncertainty

The uncertainty due to the multiplexer and demultiplexer is estimated to be between 0.9 and 1.5 ns for all the chips. This estimate is t_{valid} minus 200 ps of jitter and 300 ps of duty cycle imbalance in the clock generation circuitry. C3 in Table 7.4 shows the calculated values for each chip.

7.2.4 Design Discussion

Table 7.4 shows that the optimal bias is between 0.3 and 0.9 V at 200 MHz depending of the particular chip. The optimal valid time for all chips is in

Table 7.4 Test Results: Adder Measurements and Derivatives

	M1	S	M2	M3	M4
Chip	Valid Time of adder) at 200MHz	Bias Voltage (V)	Valid Time (bypass adder) 200 MHz	250 MHz	Process/ Design Variation
1	1.8	0.7	‡	‡	0.6
2	†	†	†	†	†
3	1.8	0.7	3.4	**	0.6
4	1.6	0.3	‡	‡	1.0
5	2.0	0.7	3.6	1.8	0.2
6	1.8	0.9	3.4	**	0.6
7	2.0	0.5	3.4	1.2	1.0
8	2.2	0.7	3.4	1.4	0.2
9	1.8	0.5	3.2	1.6	1.2
10	†	†	†	†	†
11	1.8	0.3	3.4	1.4	0.8
12	2.2	0.7	3.6	1.8	0.4
13	1.4	0.5	3.0	**	1.0
14	1.4	0.5	3.2	**	1.2
15	2.2	0.9	3.2	**	0.4
16	2.2	0.5	3.4	1.4	0.6
17	†	†	†	†	†
18	1.6	0.5	3.2	1.6	1.4
19	*	*	3.4	**	1.0
20	1.4	0.7	3.2	**	0.8

Legend:
All time components are in *ns*
† failed at 10 MHz
‡ failed due to unsealed package and broken bond wires
* fails at 200 MHz
** fails at 250 MHz
Mi = measured
S = input

Table 7.4 *(continued)*
Test Results: Adder Measurements and Derivatives

	C1	C2	C3	C4	C5
Chip	Δ Delay of Adder	Δ Delay of on-chip Tester	Δ Delay of Mux/ Demux	Adjusted t_{valid} for Adder	Data Dependent Delay
1	‡	‡	‡	‡	‡
2	†	†	†	†	†
3	1.6	1.6	1.1	3.4	1.0
4	‡	‡	‡	‡	‡
5	1.6	1.4	0.9	3.4	1.4
6	1.6	1.6	1.1	3.4	1.0
7	1.4	1.6	1.1	3.6	0.6
8	1.2	1.6	1.1	3.8	1.0
9	1.4	1.8	1.3	3.6	0.2
10	†	†	†	†	†
11	1.6	1.6	1.1	3.4	0.4
12	1.4	1.4	0.9	3.6	1.0
13	1.6	2.0	1.5	3.4	0.6
14	1.8	1.8	1.3	3.2	0.6
15	1.0	1.8	1.3	4.0	0.6
16	1.2	1.6	1.1	3.8	0.6
17	†	†	†	†	†
18	1.6	1.8	1.3	3.4	0.2
19	*	1.6	1.1	*	*
20	1.8	1.8	1.3	3.2	1.0

Legend:
All time components are in *ns*
† failed at 10 MHz
‡ failed due to unsealed package and broken bond wires
* fails at 200 MHz
Ci = calculated
C1 = M2 - M1 C2 = 5 ns - M2
C3 = C2 - 500 ps for clocking effects C4 = 5 ns - C1
C5 = C1 - M4 for estimate of the timing variations due to data dependent delays

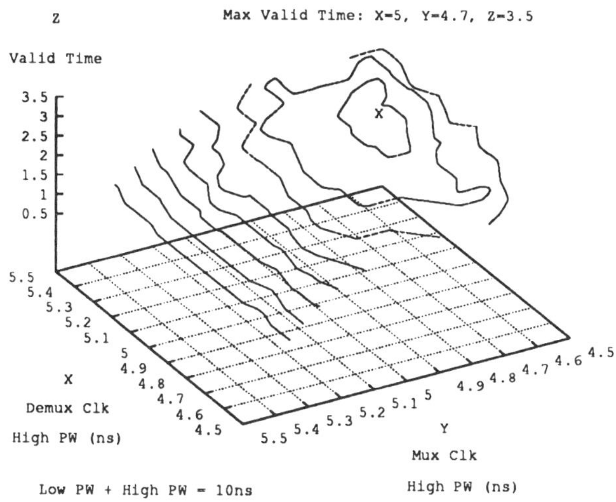

Figure 7.12 Duty Cycle Plot

the range of $1.4ns \leq \mathbf{t_{valid}} \leq 2.2ns$. This range of bias voltages shows the feasibility of post-fabrication tuning of the bias voltage to adjust the relative strengths of the PMOS and NMOS devices and thereby optimize performance.

The shmoo plot in Figure 7.10 shows a tolerance of ± 0.2 V in Vb and skew of ± 1 ns for a typical chip. The shmoo plot of Figure 7.11 shows a tolerance of ± 0.2 V in Vdd and skew of ± 1 ns when Vb is fixed. This indicates that, in order to operate the adder at 200 MHz with high reliability, it becomes necessary to have a stable power supply and bias voltage or to have a self-adjusting skew and bias in order to alleviate the Vdd fluctuation. On the other hand, reducing the operation clock rate provides increased margin for variability in Vdd. The shmoo plot of the constraint space 7.9 shows that the valid time, $\mathbf{t_{valid}}$, grows linearly with cycle time, and it is reasonable to assume that the operating region (white cell) in the shmoo plots of the Vdd sensitivity also grows linearly. Accordingly, it is estimated that the tolerance of Vdd at ± 10% requires that the adder must be operated less than 140 MHz, which still gives about five times speed up.

Data dependent delay variation was a major concern in the design of the adder circuit. Interestingly, Table 7.4 shows that there is no strong correlation between *process variation* (M4) and *valid time* (M1). In other words, smaller process variation does not necessarily imply a larger valid time. Since, M1=M2-C5-M4 and M2 is relatively constant, there is apparently some interaction between data dependent delay (C5) and process variation (M4). The effects of the process variation on the circuit is more complicated than can be easily quantified and, thus, emphasizes the need for timing analysis over the entire process space as discussed in Section 5.8.2 of Chapter 5.

Applying random test vectors, simulations show almost constant power dissipation with the observed maximum variation in power being around five percent. Since, the transient time between power fluctuations is much shorter than the thermal time constant of typical IC packages, it is expected that the chip temperature will remain relatively constant which will limit temperature variations in circuit parameters. Testing recorded an average power of 3 watts. Instantaneous power dissipation was not measured.

7.3 SAMPLER

The second design example given in this chapter is a circuit for high resolution digital sampling of an arbitrary waveform [35, 100, 102]. This circuitry has been fabricated in 1.2 μm CMOS technology and test results show a data bandwidth of up to 1 Gbit/s and a sampling resolution externally adjustable between 25 ps and 250 ps. While this design is not wave pipelined in the classical sense, parts of the circuit contain multiple "waves" of data during normal operation. Also, this circuit demonstrates several important concepts relating to the absolute limits of wave pipelined operation that were not observed in the adder.

The very fine sampling resolution is achieved by simultaneously propagating both data and clock signals through delay elements in such a way that resolution is controlled by the relative delay of clock and data signals rather than intrinsic gate speed. This is in contrast to sampling circuit techniques [45, 5] in which only the clock (or data) is propagated through delay units limiting the resolution to the intrinsic minimal gate delay. The difference in delay and thus the sampling resolution of the circuitry presented here can be very small, limited mainly by random variations in fabrication, jitter, and noise. This offers the potential to achieve unusually fine resolution for real-time sampling without the need for a high speed clock. This technique is similar to a technique proposed by Rahkonen et. al. for precise measurement of timing differences between two events [80].

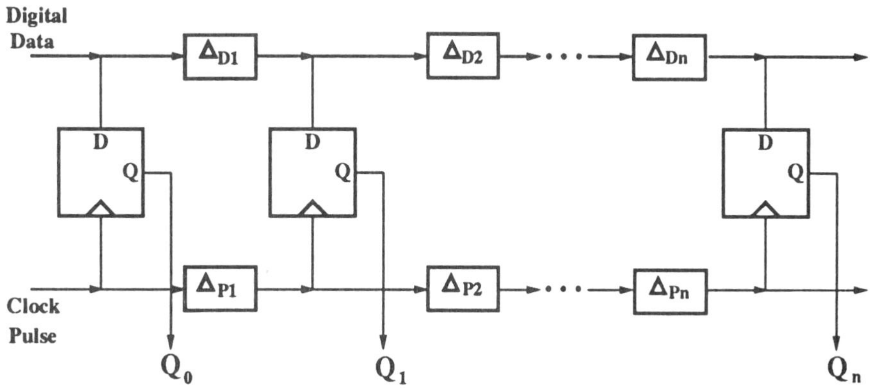

Figure 7.13 High Speed Digital Sampler

7.3.1 Design

A block diagram of the sampler is shown in Figure 7.13. As shown, data and clock are propagated through delay units (Δ_{Di} and Δ_{Pi}). Each latch samples local data with a local clock. Adjacent latches (i and $i-1$) thus sample the digital data at an effective interval of $\Delta_{Pi} - \Delta_{Di}$ as shown in Figure 7.14. Mathematically, if $D_i(t)$ is the value of the data signal at node i at time t and Q_i is the output of latch i,

$$Q_{i-1} = D_{i-1}(T) \tag{9}$$
$$Q_i = D_i(T + \Delta_{Pi}) \tag{10}$$
$$Q_i = D_{i-1}(T + \Delta_{Pi} - \Delta_{Di}). \tag{11}$$

Thus the effective sampling interval is

$$\Delta_{PDi} = \Delta_{Pi} - \Delta_{Di}. \tag{12}$$

If all $\Delta_{PDi} > 0$, the digital data is sampled at monotonically increasing points in time. If all Δ_{PDi} are equal, the digital data is sampled uniformly at this resolution. In reality, however, Δ_{PDi} will vary from stage to stage due to process and environmental variation. This must be allowed for in the specification of the maximum resolution or in some way compensated for after fabrication. In the current design, this variation is characterized and considered in the specification of the maximum resolution.

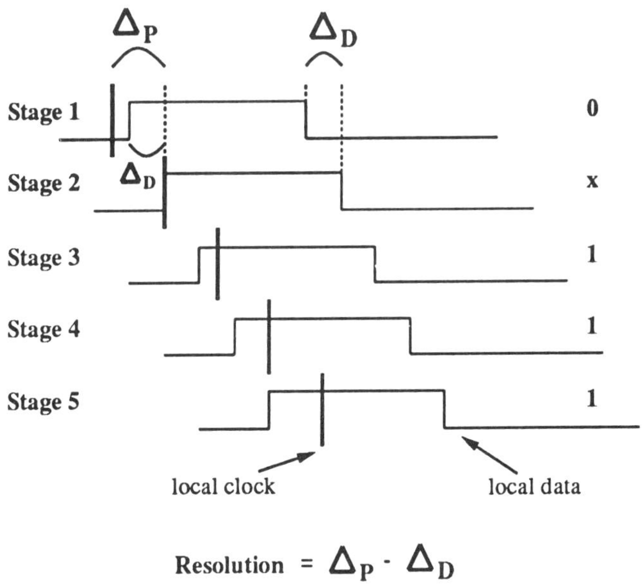

Figure 7.14 Sampler Operation

In normal operation of the implemented sampler, high speed data is propagated down the data chain at bit rates of up to 1 Gbit/s (1 ns pulse widths). Since the total propagation delay of the data chain of the implemented 64-stage sampler is greater than 12 ns, multiple data are present at various locations at any given time, and the data chain is effectively wave pipelined. The data flow through the clock and data chains is shown graphically in the logic depth-timing diagram of Figure 7.15. This figure shows the relationship between data and clock edges at each stage of the sampler. The small vertical lines indicate times at which the data is sampled at individual registers. This figure depicts fine sampling resolution relative to data edges as well as the existence of multiple data edges in the data chain at any given time.

Relating the wave pipelined data chain to the wave pipelining theory of Chapter 2, the bit data period of the sampler (defined here as T_{data}) corresponds to the clock period (T_{clk}) of a standard wave pipelined system since this is the time period at which new data is initiated into the data chain. Thus, wave pipelining theory can be used to predict the minimum data period or maximum data bandwidth of the sampler.

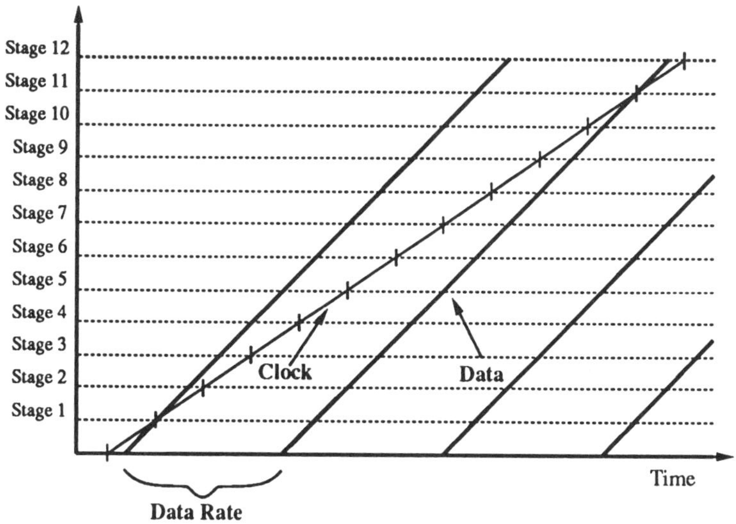

Figure 7.15 Logic Depth-Timing Diagram

From Chapter 2, the two sets of constraints for correct operation of a wave pipelined system are the register constraints and the internal node constraints. In a wave pipelined system, the register constraints specify the valid times for clocking of the output register. In normal operation of the sampler, it is valid to sample data at any point relative to the clock. Thus the register constraints are automatically satisfied. However, the internal node constraints do imply a limitation on the maximal data rate. To illustrate this limitation, Constraints 5 of Chapter 2 are rewritten below for the sampler.

$$T_{data} + t_{min}(i) \geq t_{max}(i) + t_{stable}(i+1), \forall i \quad (13)$$

$$T_{data} \geq t_{max}(i) - t_{min}(i) + t_{stable}(i+1), \forall i \quad (14)$$

In these constraints, $t_{max}(i)$ refers to the maximum time that it takes data to travel down to the data chain to the output of data delay element i. $t_{min}(i)$ refers to the minimum time. Since there is only one physical path through this delay chain, the only difference in $t_{max}(i)$ and $t_{min}(i)$ is due to dynamic variations such as power supply fluctuations, bias voltage fluctuations, signal coupling noise, and other sources of jitter. As explained in Chapter 2, $t_{stable}(i)$ relates to the amount of time that the input to data delay element i must be stable in order to the output of the element to fully switch. For the single stage differential inverter data delay units discussed in the next section, this relates to the intrinsic switching speed of the gate. In general for current technologies and careful design, $t_{stable}(i+1)$ will be greater than $t_{max}(i)$-$t_{min}(i)$ and will be the dominate limiting factor of the sampler data rate.

The maximum resolution of the sampler relates to the ability of the designer to accurately match or control the delays of the clock and data chains. This is similar to the concept of delay balancing for wave pipelined circuits although for a different purpose. Therefore, many of the techniques described in Chapter 6 used to match circuit delays for wave pipelined operation can be used to increase the sampling resolution of this circuit. Since the sampler clock and data components to be matched are similar and will be in close physical proximity, this demonstrates, in some sense, a limit to the degree of matching that is possible in general wave pipelined systems.

7.3.2 Implementation

As mentioned above, there are three main components of the sampler architecture. These are a clock delay unit, data delay unit, and a sampling latch. In this section, design considerations and circuit implementation are discussed for each of these components. Also, circuitry for high speed testing of the sampler is presented along with considerations for physical design and layout.

The fabricated test chip contains 64 sampling stages. As shown in Figure 7.16, a differential bistable latch was designed to sample data from the delay units. To drive the differential data inputs of the latch, a delay unit consisting of a differential CMOS inverter whose delay is controlled mainly by bias voltage Vb_Dp was used. To drive the single ended clock input of the latch, a delay unit was used that consisted of two biased CMOS inverters whose delay is controlled by a bias voltage Vb_C.

Bistable Latch

The circuit diagram of the bistable sampling latch is shown in Figure 7.17. As shown, the latch samples the differential data, in and inb, with a single ended clock, Cst. While Cst is high, the latch is transparent with Q and Qb following in and inb. The data is thus latched on the falling edge of the clock. The primary considerations in the design of the bistable latch were metastability,

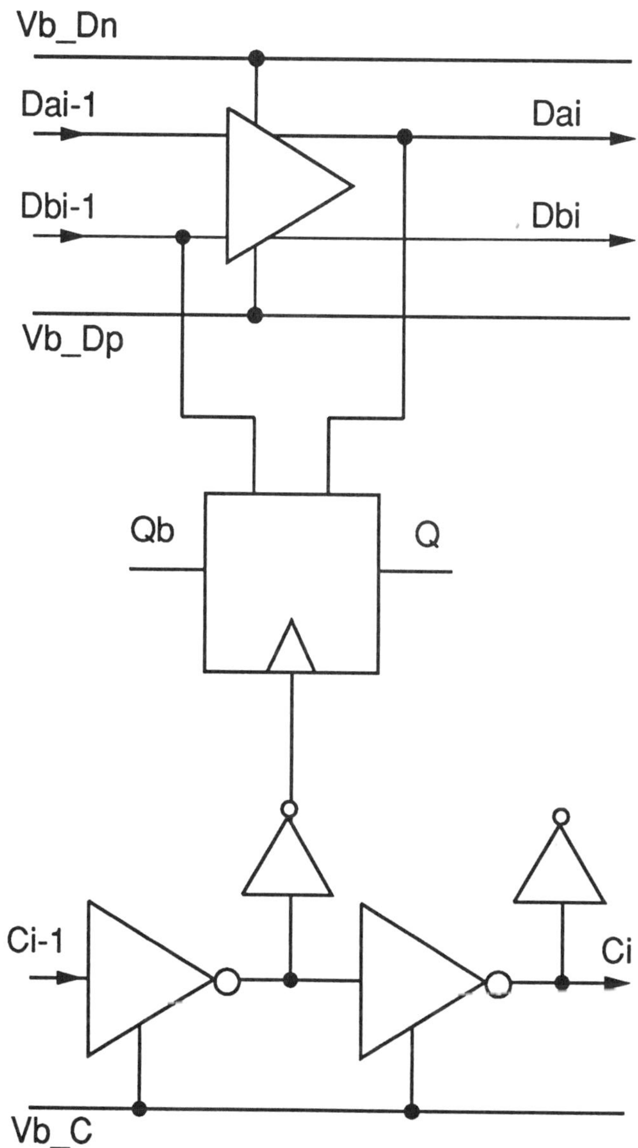

Figure 7.16 One Stage of Sampler Circuit

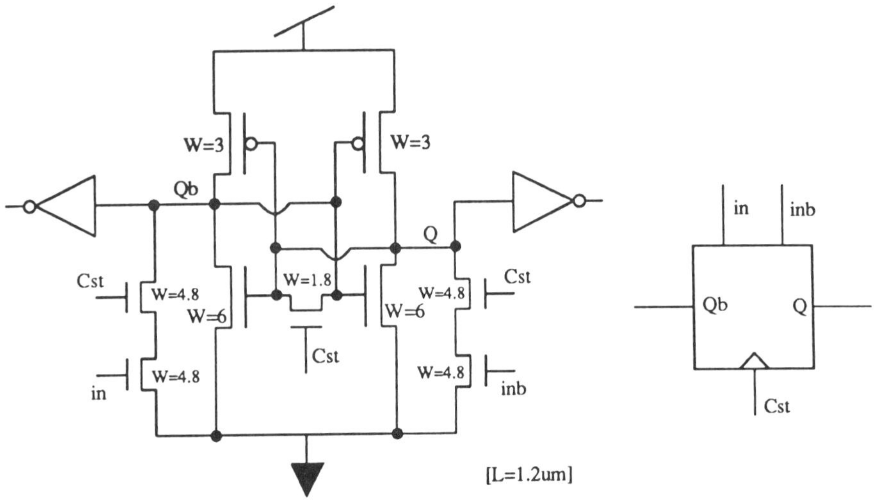

Figure 7.17 Cross Coupled Bistable Latch

independence of sampled data from the previous state of the latch, and power consumption.

First, metastability is of concern since data may easily be sampled during transition. To reduce potential problems in this area, a ratio of Wn/Wp=2 was chosen for the cross coupled inverters [85, 46]. This ratio has the effect of decreasing the metastable resolving time. Results from CAzM [23] simulation of the latch (extracted from layout) showing output resolving times are in Table 7.5. Two cases are shown. The first has both inputs equal at the midpoint of the differential input data swing (in=inb=3V). The second starts with an initial difference between the differential inputs of 0.1mV. Times at which the difference in output is 0.5V and 4V are shown. Resolving time for the two cases are thus 4.82ns and 2.48ns respectively. Note that it is also important to have the output capacitance of the bistable inverters well matched to insure correct settling of the latch when both inputs are close to each other [101].

It should be noted that although metastability is important, the requirements are less stringent for this sampling methodology than for the conventional sampling method utilizing a high speed clock and latch. Since the data is, in effect, converted from serial to parallel, the frequency at which the clock is applied to individual latches, for continuous sampling of the input data, can be relatively slow. This typically gives ample time for resolution from the metastable

Table 7.5 Metastability Analysis of the Bistable Latch

Initial Δ inputs	Resolving Time Q-Qb=0.5V	Resolving Time Q-Qb=4V
0.0 mV	4.09 ns	4.82 ns
0.1 mV	1.73 ns	2.48 ns

state. However, it is still important to consider the metastable time to insure that these requirements are met. Also, circuit noise while the latch is in the metastable state will potentially affect the final latched state. These concerns indicate that a reasonably fast resolution time is needed.

The second major concern is that sampled data should be independent of the previous state of the latch. In the transparent latch used here, this problem can arise if the latch outputs do not follow the data fast enough during the transparent phase. This can cause the latched output to be dependent on the preceding data. To increase the speed of the latch and thus reduce any "memory effect", a 1.8 μm transistor was added between the differential outputs Q and Qb. This transistor, in effect, reduces the signal swing to about 0.9V at the latch outputs when the clock, Cst, is high. This reduces the effect of the previous state of the latch and allows the latch outputs to follow short input data pulses. In addition, the sampled data should be independent of data arriving after the clock edge. This dependence was minimized by placing the transistors clocked by Cst between the input transistors (in and inb) and the output nodes (Qb and Q) and by minimizing interconnect coupling between inputs in and inb and outputs Q and Qb.

Finally, power consumption of the latch was a concern since the latch will conduct current from Vdd to ground during the transparent phase (Cst=5V) both during switching and while the data is stable. Figures 7.18 and 7.19 show the current paths when data is stable (in=5V, inb=0.9V) and when data is in transition (in=3V, inb=3V). These figures show results of CAzM simulations under nominal process parameters. As shown, peak power supply current is approximately 540μA when Cst=5V and data is in transition and 440μA when Cst=5V and data is stable. Current is less than 0.2μA in the stable latched state (Cst=0V). Note that the inverter ratio of Wn/Wp=2, chosen for metastability concerns, has an added benefit in that the current, when data is in transition, will be reduced. This is due to the fact that the widths of the PMOS transistors

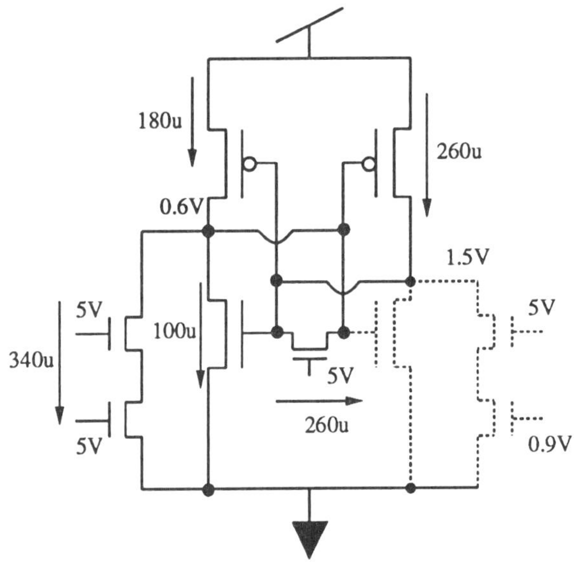

Figure 7.18 Cross Coupled Latch Operation (Cst=5V, in=5V, inb=0.9V)

are reduced and the Q and Qb levels are shifted such that the two NMOS transistors of the latch are OFF under these conditions (see Figure 7.19).

Delay Units

The primary considerations in the design of the data and clock delay units were controllability of delay, low noise, and manageable power. To drive the differential data inputs of the latch, a differential inverter (Figure 7.20) was designed for the data delay unit. The voltage swing of the differential inverter is from about 0.9V to 5V and the delay is controllable mainly by bias voltage Vb_Dp. As shown in Figure 7.21, the delay of this inverter is controllable nominally from about 200 ps to 350 ps. An additional consideration in the design of the differential inverter was the ability to pass high speed data (up to 1 Gbit/s). This implies the need for high differential gain and high slew rates at high frequencies. This in turn meant an increase in current and power.

Since only a single ended clock was needed for the latch, a biased CMOS inverter (Figure 7.22) was designed for the clock delay unit. Each clock delay unit consists of two biased CMOS inverters whose delay is controllable by a bias

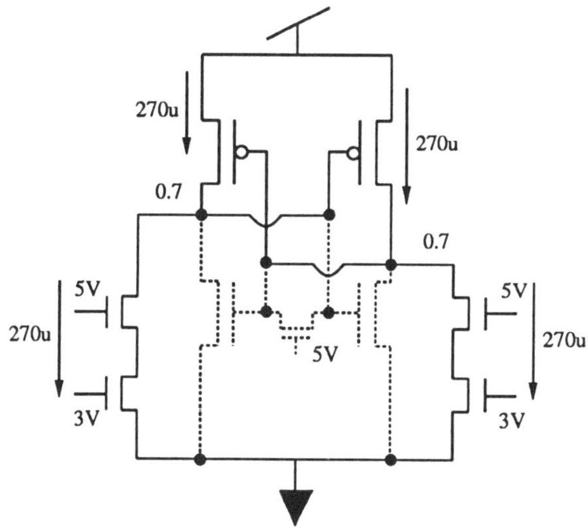

Figure 7.19 Cross Coupled Latch Operation (Cst=5V, in=3V, inb=3V)

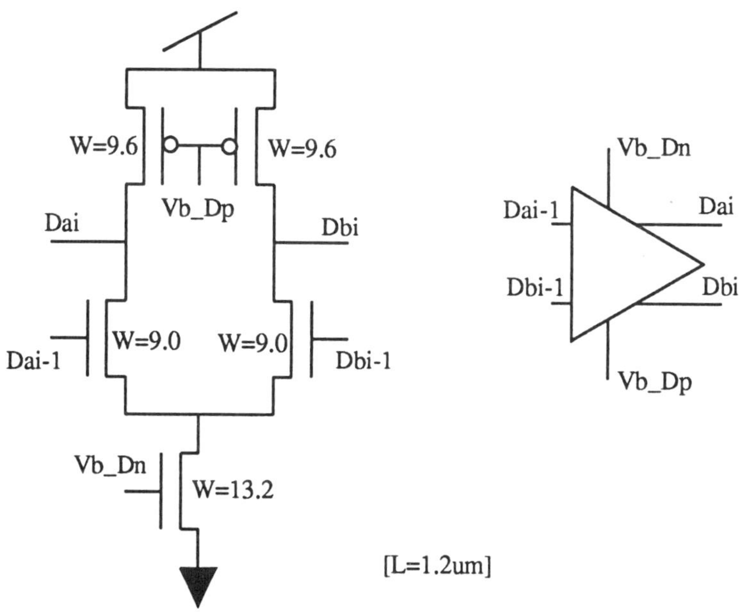

Figure 7.20 Differential Inverter Data Delay Unit

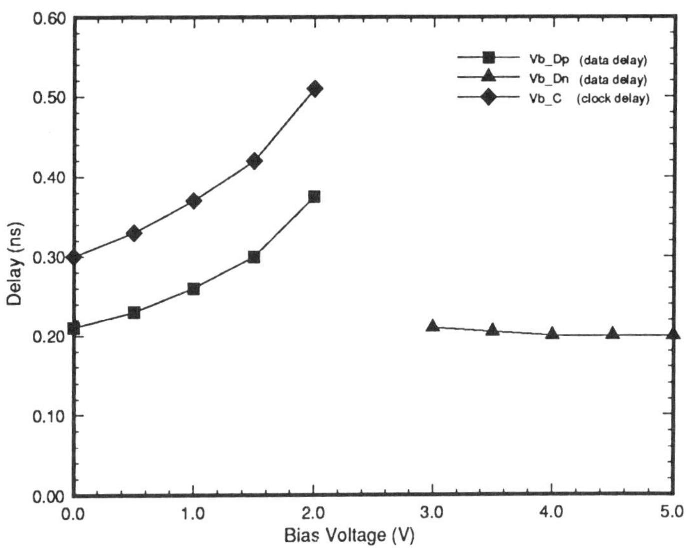

Figure 7.21 Delay vs. Bias Voltage for Data and Clock Delay Units

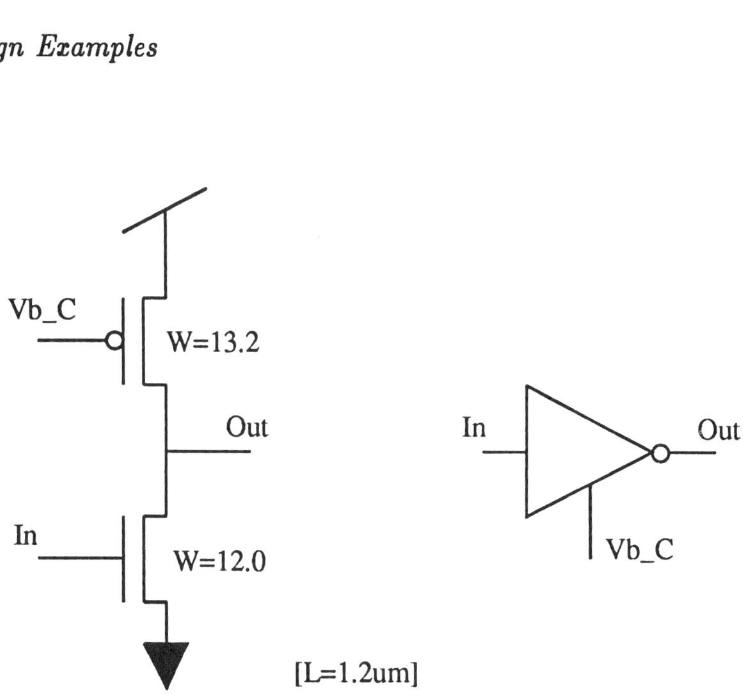

Figure 7.22 Biased CMOS Inverter Clock Delay Unit

voltage Vb_C from about 300 ps to 500 ps as shown in Figure 7.21. As shown in Figure 7.16, standard static inverters are used between the clock delay units and the latch inputs to drive the clock input fully to ground.

The peak current (during switching) of the differential inverter is about 1.7mA with a standby current of 1.1mA. The current of the two-inverter clock delay unit is relatively constant at about 1.3mA since exactly one inverter is conducting during standby and currents tend to balance each other during switching. While these biased CMOS and differential inverters consume more power than static CMOS inverters due to static power dissipation, switching noise due to transient current spikes is greatly reduced [45, 24, 64].

Test Structure

Due to a lack of high speed test equipment and the difficulties of high speed I/O buffers in 1.2 μm CMOS for 1 Gbit/s data signals, high speed test capabilities were designed on-chip. An eight input XOR tree (Figure 7.23) was designed to multiplex slow speed data signals to generate a high speed signal for the sampling circuit [117, 116]. Since any change of one input to the XOR tree produces a corresponding change at the output, the XOR tree can be used to produce very short data pulses at its output by precise control of input skews.

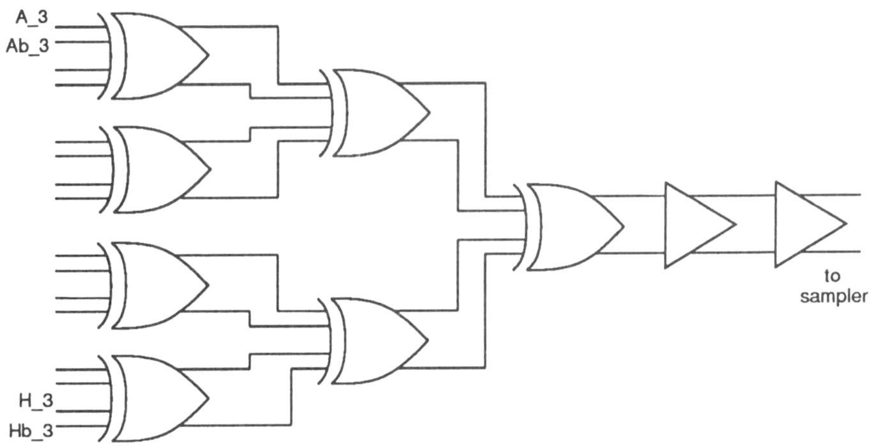

Figure 7.23 Eight-Input XOR Circuit

To achieve the necessary output data pulse widths, the XOR tree was implemented with two input differential XOR gates (Figure 7.24) that allowed for pulses as small as 1 ns to be input to the sampling circuit. Pull down paths are cross-coupled to reduce input dependent delay variation. Note that, in this application, since inputs to the XOR tree are skewed relative to each other, simultaneous vs. single transition delay variation, as described in Chapter 6, is not a problem. Transistor sizes are designed such that output levels of the XOR tree are similar to the differential inverters of the sampler. Furthermore, as shown in Figure 7.23, the outputs of the XOR tree are passed through two differential inverters to better align differential signals and match signal levels.

Inputs to the chip are through standard MOSIS single-ended digital pads. Differential inputs are generated using the single-ended to differential converter shown in Figure 7.25. The outputs of the two inverter/three inverter conversion circuit are passed through two differential inverters to better align differential signals. For reading data from the sampler off chip, a standard two-phase slow-speed 64-bit shift register is included. This shift register is loaded from the sampling latches in parallel and read off-chip serially.

Layout

A test chip containing the 64-stage sampler along with the XOR multiplexer was implemented in the MOSIS 1.2μm CMOS process. The floorplan of the

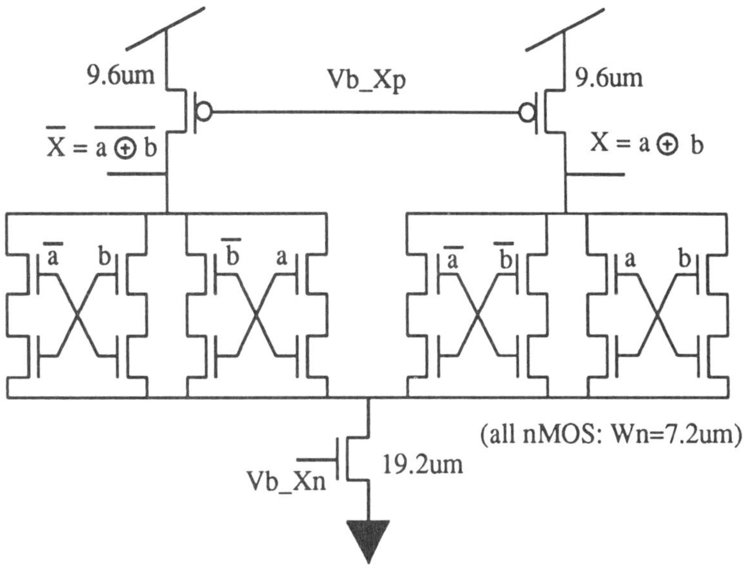

Figure 7.24 Two-Input Differential XOR

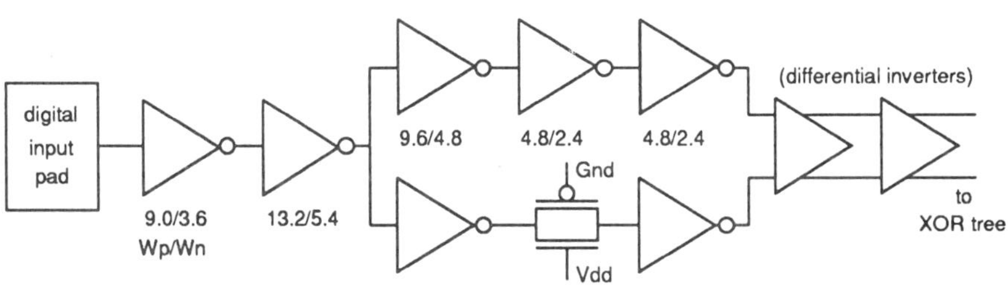

Figure 7.25 Single to Differential Input Structure

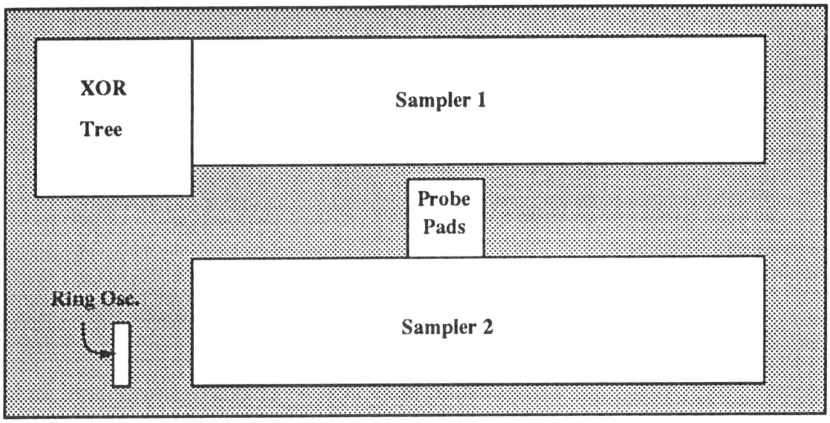

Figure 7.26 Sampler Floorplan

chip is shown in Figure 7.26. This chip contains two versions of the sampler structure. One version is fed by the XOR multiplexer and the input to the other is taken directly from chip probe pads for future direct testing of the sampler. Also, as shown in the floorplan, a 37-stage ring oscillator consisting of standard static inverters (Wp/Wn = $6\mu m/3\mu m$) was included as a rough measure of process variation. The ring oscillator could be turned on or off for variation in digital switching noise on-chip. As with the adder, MAGIC [65] was used for layout and CAzM [23] was used for circuit simulation.

To reduce differences in delay due to variations in loading, careful balancing of interconnect was done in the sampler and in the XOR tree. To achieve maximal process tracking of interconnect capacitance, this balancing was done categorically as with the adder (Section 7.2.2). Note, however, that in this case the area and power overhead was quite low since there were no long interconnect lines. Clock delay unit and data delay units for each stage were placed in close physical proximity to minimize delay variation due to gradient process variation.

For highest performance, it is very important to reduce all sources of noise. This includes crosstalk between signal lines, switching noise coupled to bias lines, and variations in power supply levels. Using techniques similar to those described in Section 6.5 and as used in the adder, a ground line is routed between long parallel signal lines, and bias lines (Vb_Dn, Vb_Dp, Vb_C, etc.) are shielded above. Also, many decoupling capacitors are added between power and ground lines to keep voltages more stable. Separate power supplies are provided for the input pads and for the ring oscillator.

A photomicrograph of the fabricated chip is shown in Figure 7.27.

7.3.3 Test Results

To verify the fabricated sampler circuit, experimental tests on the fabricated device were performed using a HP82000 chip tester at the Microelectronics Center of North Carolina (MCNC). Tests were included to determine the amount of process variation and jitter affecting the device performance, the effects of bias voltages on sampling resolution, and the maximum bandwidth of the sampling system. Forty chips were received from MOSIS and 39 functioned correctly sampling a 1.5 ns pulse at about 150 ps resolution.

To determine the effect of process variation and jitter on the sampling operation, data transitions were applied at the inputs to the XOR tree to produce successive data pulses of 1.5 ns, 2.0 ns, and 2.5 ns at the data input of the sampler. By changing the time of the transition of the clock input relative to the data inputs, the three data pulses are sampled at different locations in the sampler. Then, by observing the relative locations of the edges, a measure of the amount of the uniformity of sampling resolution in different parts of the sampler and the amount of jitter is obtained. Figure 7.28 shows the results of this test when the clock transition edge is skewed over a range of 2 ns to 9 ns in steps of 50 ps (the edge placement resolution of the tester). At each clock skew point, the test was repeated 100 times to get the average location of each edge. Nominal bias voltages of Vb_Dn=5V, Vb_Dp=0V, Vb_C=0.5V and Vdd=5V were used for this test. As with any sampling strategy in which data and clock are not synchronized, the observed number of samples per pulse can vary by at most one depending on the relative alignment of data and clock when the data pulse width is not an exact multiple of the sampling resolution as shown in Figure 7.29. In addition, when sampling at a data edge, jitter of less than the signal transition time will cause an uncertainty in the least significant bit of the sampler location. Thus, when both edges of a pulse coincide with sampling times, a variation of up to two samples per pulse is possible (Figure 7.30). For the 1.5 ns pulse in the above test, the observed average number of samples varies from 11.0 to 12.9 for different locations in the sampler, indicating a small amount of process variation and random jitter.

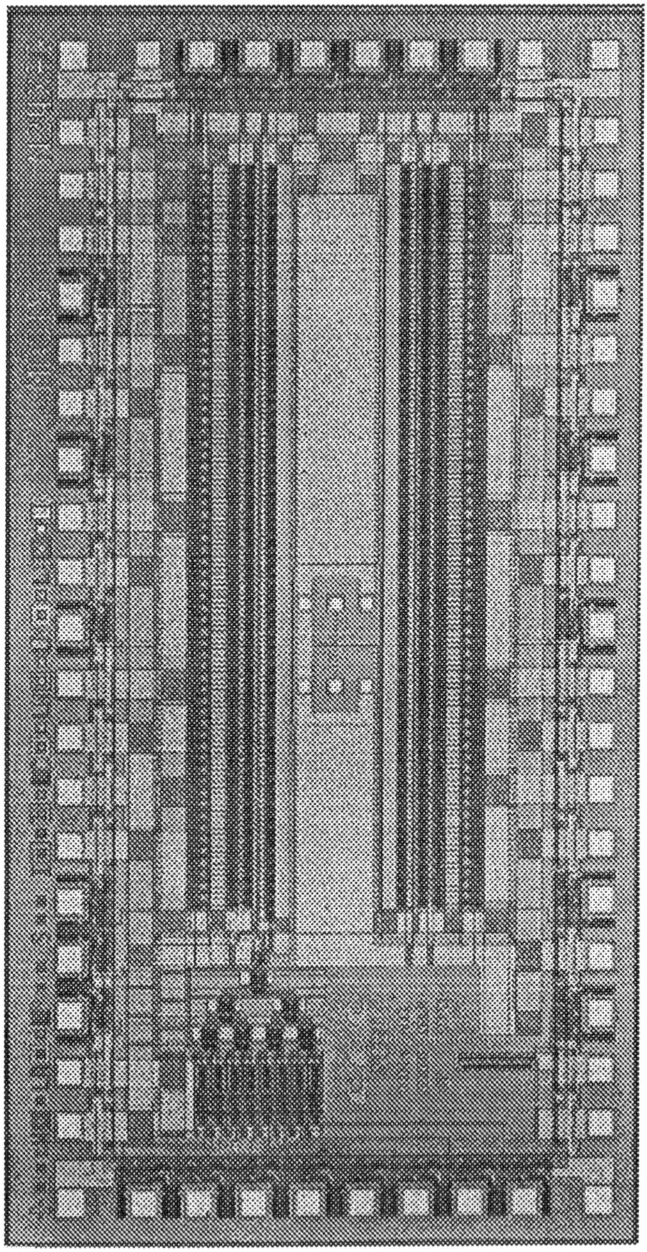

Figure 7.27 Photomicrograph of Sampler Test Chip

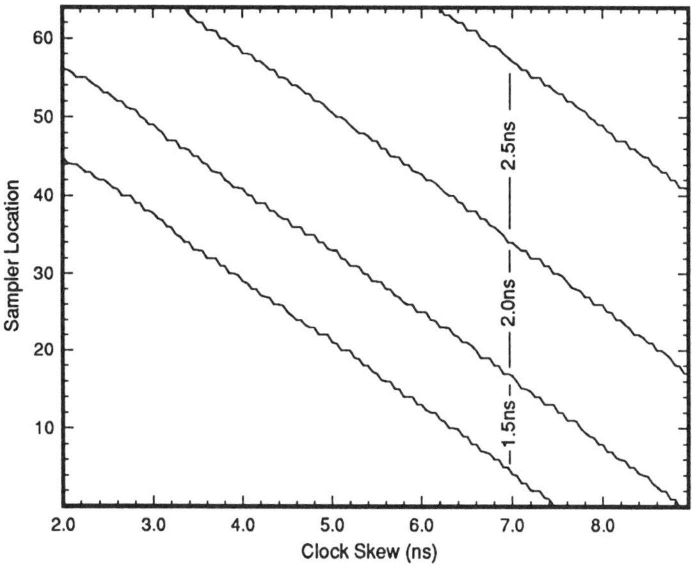

Figure 7.28 Edge Location vs. Clock Skew

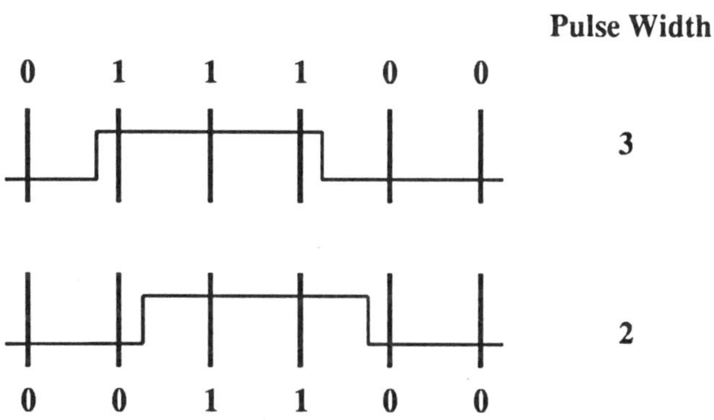

Figure 7.29 Unknown Relationship Between Clock and Data

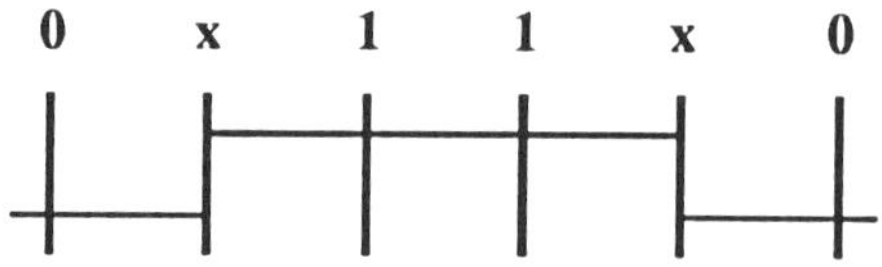

						Pulse Width
0	0	1	1	0	0	2
0	0	1	1	1	0	3
0	1	1	1	0	0	3
0	1	1	1	1	0	4

Figure 7.30 Sampling at an Edge

To determine the effect of bias voltages on sampling resolution, each of the three biases, Vb_Dn, Vb_Dp and Vb_C, and the power supply, Vdd, were varied in turn while keeping the other biases at nominal values. Figure 7.31 shows the results of this test when a 2 ns pulse is applyed to the input of the sampler. This test shows clearly that the PMOS biases Vb_C and Vb_Dp have the most effect on the sampling resolution and shows a range of resolution from 40 ps to 230 ps as a result of changes in bias voltages. Note also that the sampling resolution is relatively insensitive to the power supply voltage. Over a range of $\pm$ 10% in power supply, the sampling resolution varies by only $\pm$ 4%. This is due to the delay tracking of the differential inverter and the biased CMOS inverter under power supply variations. *Resolution limits* were obtained by setting Vb_C = -0.5, Vb_Dp = 1.26 for a resolution of 25 ps and by setting Vb_C = 1.5, Vb_Dp = 0 for a resolution of 250 ps as shown in Figure 7.32.

To determine the effects of digital noise on the operation of the sampler, a jitter test was run with the sampler set for very high resolution sampling. For this test, Vb_C = -0.5V, Vb_Dp = 1.26V, and Vb_Dn = 5V for a resolution of about 25 ps. A 1.5 ns pulse was applied to the data input and the number of samples for the pulse was recorded. This test was conducted 1000 times over about 20 minutes and Figure 7.33 shows statistics of the number of samples that were recorded. As described above and shown in Figure 7.30, a sampled pulse width can vary over three values (2,3,4 in Figure 7.30) due to uncertainty at the data edges. This corresponds to the widths of 58, 59, and 60 for the jitter test of Figure 7.33. The widths outside this range (57, 61, and 62) indicate

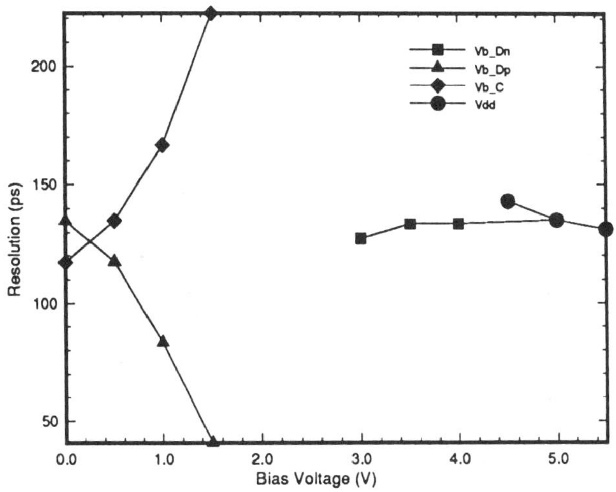

Figure 7.31 Resolution vs. Bias Voltage

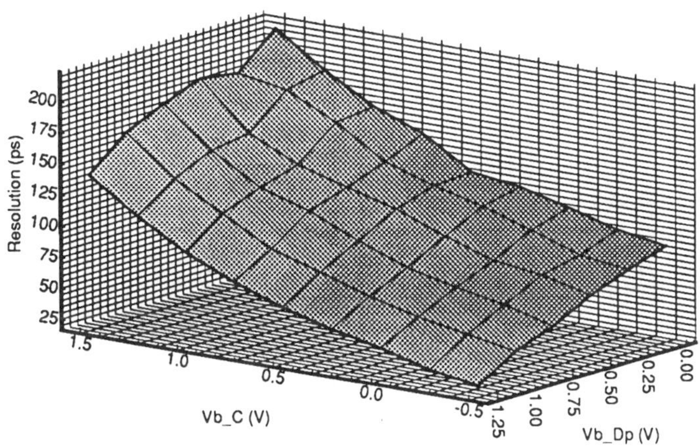

Figure 7.32 Resolution vs. Bias Voltages

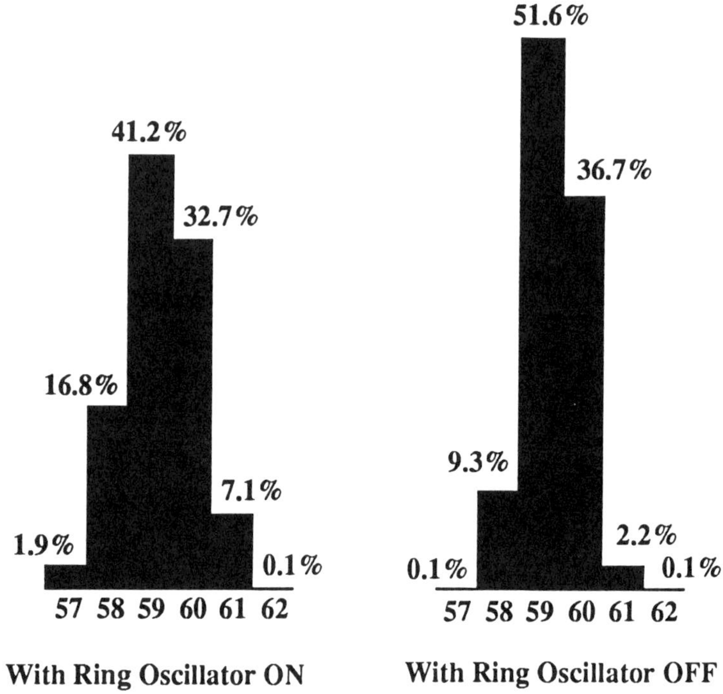

Figure 7.33 Jitter with/without Ring Oscillator

jitter between ±25ps and ±50ps at each data edge due to the combined effects of the the HP tester and the sampling unit. This is shown conceptually in Figure 7.34. Also, it can be seen from Figure 7.33 that the variation in sampled pulse width is somewhat less when the ring oscillator is turned off. Clearly, the noise produced by the ring oscillator affects the jitter in the sampler even though the ring oscillator is physically located on the opposite side of the chip. This again emphasizes the need for tight control of all noise sources.

Finally, to determine the maximum useful bandwidth of the sampling unit, the difference in skews at the inputs to the XOR tree was successively reduced to thereby produce successively shorter pulses at the input of the sampler. This process was continued until the pulse was no longer observed in the sampler. Using nominal bias voltages, correct operation was observed down to a 1 ns input pulse width. Note that this limit may have been due to a limitation of the XOR tree since internal circuit nodes were unobservable with equipment

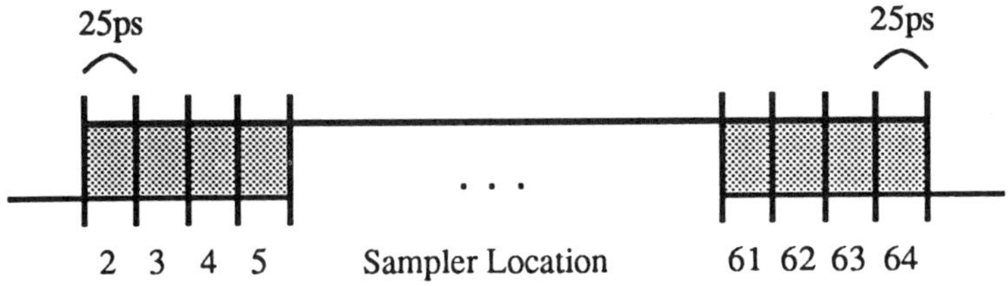

Figure 7.34 Jitter Test (1.5ns pulse at 25ps resolution)

Table 7.6 Design and Performance Statistics: Sampler

Transistors (sampler)	2624
Transistors (samp. + XOR)	3108
Die size	1795 μm x 3647 μm
Total I/O pins	58
Process	MOSIS 1.2μm NWell
Package	84 pin PGA
Maximum Bandwidth	1 Gbit/s
Sampling Resolution	25 ps to 250 ps
Power Dissipation	1 W

used, but this test does indicate at least a 1 Gbit/s useful bandwidth in the sampler.

Table 7.6 gives a summary of some design and performance statistics of the sampler chip.

7.4 CONCLUSION

This chapter has presented VLSI implementations of two designs demonstrating concepts of wave pipelining. The first design was a fully wave pipelined sixteen-bit adder circuit that operated at a maximum rate of 250 million operations per second in 2μm CMOS. This chip demonstrated many practical aspects of wave pipelined design and confirmed the theoretical register constraint bounds developed in Chapter 2.

The second example was the implementation of a high-resolution digital sampling circuit. This design had a maximal bandwidth of 1 Gbit/s and a minimum sampling resolution of 25 ps in 1.2μm CMOS. This design demonstrated practical limits of wave pipelining when circuit delays are exactly equalized. The limitations imposed by fundamental gate delays and by variations due to noise, jitter, and process were discussed.

8

CONCLUSIONS

This monograph has concentrated on four important considerations in the design of working wave pipelined systems:

- theoretical development of the constraints for correct clocking,
- detailed timing analysis of combinational logic blocks,
- practical design concepts, and
- actual design examples.

Through a rigorous theoretical development of the constraints for correct clocking for single stage and multiple stage systems, important new generalized results have been presented relating to performance limits of wave pipelined circuits. These results have included the identification and specification of distinct and disjoint regions of valid operation whose sizes are dependent on the clock period, intentional clock skew, and the global clock latency. An analogous relationship between the multistage system and the single stage system in terms of performance limits was also developed. New results were shown to be consistent with previous special case results and examples were given demonstrating the applicability of the formulation.

To support the theoretical formulation, algorithms and methodology for detailed exact timing analysis of combination logic blocks were presented. This timing analysis considers the effects of input gate ordering, gate delay dependency, rising vs. falling delay dependency, and the delay dependence on relative skew between multiple input transitions particularly important for the analysis of high performance CMOS circuits. Although this problem was shown to be

NP-complete, algorithms and heuristics were developed and implemented in a prototype timing analyzer, *XTV*, that was shown to be feasible for moderate size circuits. Results from the application of these algorithms on a set of benchmark circuits demonstrated the inaccuracies of less detailed gate models of previous approaches and the need for this level of accurate timing analysis.

Next, practical considerations of wave pipelining were discussed. These considerations included the choice of architecture, reduction of path delay variation, choice of basic logic gates, the effects of process and environmental variations, and physical design.

Finally the concepts and formulations developed were put into practice with the description of two actual wave pipelined designs. The first was a 250 MHz wave pipelined 16-bit adder in fabricated in MOSIS 2μm CMOS and the second was a digital sampling circuit with 1 GBit/s bandwidth and 25 ps resolution fabricated in MOSIS 1.2μm CMOS. For both examples, an overview of the architecture was presented along with important design considerations and test results. These examples demonstrate the validity of the proposed theoretical and design methodologies and show the feasibility of wave pipelining using the CMOS technology.

In any research area as broad as the area of wave pipelining, there are always many possible extensions and directions for future work. First, the optimization methods developed in Chapters 2 and 3 could be extended. The formulation of this monograph optimized clock skews assuming fixed circuit blocks and register positions. It has been shown that similar concepts such data path delay equalization [107, 22] and retiming [57, 56] can also be used to increase circuit performance. It would be interesting to investigate optimization by all of these methods together under possible additional constraints such as area, power, and perhaps tolerance to process/environmental variation.

It would also be interesting to investigate the implementation of multistage circuits with feedback employing wave pipelining in some or all stages. There is inherently a tradeoff between the addition of physical registers and the allowance for additional wave pipelining. This tradeoff should be studied more and perhaps included in the optimization process along with retiming. In this case, the procedure would not only determine the position of registers but also whether registers are needed at all. This would lead to an generalized method of synthesizing circuits specifically for optimal wave pipelined performance.

In the timing analysis area, there are several extensions and improvements that could be made. Most of which are discussed in detail in the latter part of

Chapter 5. To summarize, extensions could be made to include more accurate delay models that capture the effects of slope and load dependent delays. Also, delay ranges or distributions could be implemented to insure accurate bounds on delay over the range of process/environmental variation. In order to analyze larger circuits, methods for increasing the efficiency of the search routines and data structure could be investigated. Along with improvements to the existing algorithm, there are two properties of many large circuits that can be leveraged to reduce computational complexity. First, the hierarchical structure of many circuits can be used to partition the circuit so that detailed analysis need only be done on smaller circuit blocks. Second, the regular nature of circuits containing such structures as trees, arrays, etc. can be considered such that analysis is not duplicated for similar structures. These properties could be used to analyze much larger circuits while still maintaining the needed accuracy.

In addition to these extensions, the existing analysis framework could be used to accurately solve related problems. First, this framework could be used to accurately detect the existence of glitches at internal circuit nodes. This is important both for accurate timing analysis and for the implementation of dynamic CMOS logic structures. The second area for future extensions is the problem of accurately determining worst case dynamic power consumption in CMOS circuits. Since dynamic power consumption of CMOS circuits is related to switching behavior, it is obvious that there is a relationship between accurate determination of transition times and accurate determination of dynamic power. Techniques similar to those presented in Chapter 5 could potentially be used to search for vectors with the worst case amount of internal node switching or the worst case amount of simultaneous switching.

From a practical standpoint, one of the largest concerns with the example circuits presented in Chapter 7 is the amount of power consumption. This is primarily due to the use of the biased CMOS gates. While these gates are excellent from the standpoint of low data dependent delay variation, the gates consume static power in the low logic state. While some alternative circuits have been presented that do not consume static power, these circuits have more variation in delay due to data dependency. It would be interesting to investigate alternative logic structures that both consume less static power and have low data dependent delay variation. In addition to the power considerations, the amount of chip area devoted to balancing interconnect is of concern for circuits such as the adder example. While the conservative balancing methodology used in the adder provided for maximal tracking over process variation, it may be possible to use a less rigid matching approach if better information on process parameter distributions, better process control, and more accurate parasitic extraction are available. For example, instead of maintaining equal routing lengths, various sizes of transistors could be deployed to drive various length of routing capacitance. This would substantially reduce area penalty and power usage.

Another important practical consideration is the testability analysis of wave pipelined circuits. While wave pipelined circuits have many testability concerns that are common to any high performance circuit such as generating test vector sets guaranteed to produce the absolute earliest and latest output transitions of a particular logic block over all possible process variations, wave pipelining also creates some additional concerns such as the inability to stop testing and examine the state of the circuit. These concerns should be investigated to fully determine the commercial feasibility of aggressive wave pipelining.

Finally, concepts used in wave pipelining can be extended to a broader class of circuits. The power and performance advantages of wave pipelining are derived from the fact that performance depends on relative delays or differences in delays rather than on absolute delays. At the physical level, the performance thus depends on the ability to match or coordinate circuit delays. For single-chip integrated circuit designs, this approach also has the advantage that designs are less sensitive to process and environmental variations since delays will, to some degree, track across a single die. This concept was shown both in the adder example and the sampler example. This suggests that a more general design methodology based on delay matching, here referred to as the *Delay Coordination Method (DCM)*, could be used in a broader class of circuits. In a sense, the sampler architecture pushes this concept of delay matching to the limit and has shown correct performance to the point that noise considerations dominant. For future research, it would be interesting to investigate the use of DCM circuits, such as the sampler, for applications such as signal processing and telecommunications where performance requirements are high and pipeline latency is allowable.

A

EXAMPLE MODEL FILE

models.ta

```
nand 2
LOW  LOW   high
LOW  HIGH  high
LOW  RISE  high
LOW  FALL  high
HIGH LOW   high
HIGH HIGH  low
HIGH RISE  falling 600
HIGH FALL  rising  790
RISE LOW   high
RISE HIGH  falling 470
RISE RISE  falling 680
RISE FALL  high
FALL LOW   high
FALL HIGH  rising  590
FALL RISE  high
FALL FALL  rising 360

nor 2
LOW  LOW   high
LOW  HIGH  low
LOW  RISE  falling 720
LOW  FALL  rising  620
HIGH LOW   low
```

```
HIGH HIGH  low
HIGH RISE  low
HIGH FALL  low
RISE LOW   falling 530
RISE HIGH  low
RISE RISE  falling 360
RISE FALL  low
FALL LOW   rising  440
FALL HIGH  low
FALL RISE  low
FALL FALL  rising  680

and    2
LOW  LOW        low
LOW  HIGH       low
LOW  RISE       low
LOW  FALL       low
HIGH LOW        low
HIGH HIGH       high
HIGH RISE       rising  1084
HIGH FALL       falling 1150
RISE LOW        low
RISE HIGH       rising  900
RISE RISE       rising  1110
RISE FALL       low
FALL LOW        low
FALL HIGH       falling 950
FALL RISE       low
FALL FALL       falling 720

or 2
LOW  LOW        low
LOW  HIGH       high
LOW  RISE       rising  1150
LOW  FALL       falling 980
HIGH LOW        high
HIGH HIGH       high
HIGH RISE       high
HIGH FALL       high
RISE LOW        rising  960
RISE HIGH       high
RISE RISE       rising  790
```

```
RISE FALL        high
FALL LOW         falling 800
FALL HIGH        high
FALL RISE        high
FALL FALL        falling 1040

not  1
LOW   high
HIGH  low
RISE  falling 360
FALL        rising 430
```

B

CALCULATION OF TOLERANCE OF PARAMETRIC VARIATIONS

In this section, an expression for the amount of variation that can be tolerated is derived as described in Section 6.4.2. For this derivation, let c be the amount of variation along the data path through the combinational logic such that the maximum logic delay is ct_{max} and the minimum logic delay is ct_{min}. Also, assume that the programmable delay unit is also tracks variation along with the combinational logic. (This assumption is valid if the programmable delay unit is fabricated on the same substrate as the combinational logic.) For simplicity assume, setup time, hold time, latch delay, and unintentional clock skew are zero.

Now, consider three cases:

Case 1 : k=0

Referring to equations (1) and (2) of Chapter 2:

$$c\Delta \geq ct_{max} \tag{B.1}$$

and

$$T_{clk} \geq c(\Delta - t_{min}) \tag{B.2}$$

Notice here that T_{clk} is only bounded below. Thus, a direct tradeoff can be made between the amount of variation to be tolerated and the clock period.

Case 2 : k=1

In this case, equations (1) and (2) of Chapter 2 simplify to:

$$c\Delta \leq ct_{min} \tag{B.3}$$

and

$$T_{clk} \geq c(t_{max} - \Delta) \tag{B.4}$$

Similar to case 1, a tradeoff can be made between the amount of variation to be tolerated and the clock period.

Case 3 : $k > 1$

Referring to section 6.4.2, further assume that t_{max}=t_{min} and $\Delta = 0$. This will give an upper bound on the amount of variance that can be tolerated. Note that under these conditions, the feasible regions for each k, as shown in Figure 2.9, will be maximized thus allowing for he maximum amount of variability.

Consider a region of arbitrary k. This region, along the $\Delta = 0$ axis, is bounded below by $\frac{ct_{max}}{k}$ and above by $\frac{ct_{max}}{k-1}$. Thus a nominal $T_{clk}(t_{nom})$ must fall within these limits.

$$\frac{ct_{max}}{k} \leq t_{nom} \leq \frac{ct_{max}}{k-1} \tag{B.5}$$

rearranging terms an expression for bounds on the parameter c (c_{high} and c_{low}) is derived.

$$\frac{(k-1)t_{nom}}{t_{max}} \leq c \leq \frac{kt_{nom}}{t_{max}} \tag{B.6}$$

Define

$$c_{low} \equiv \frac{(k-1)t_{nom}}{t_{max}} \tag{B.7}$$

$$c_{high} \equiv \frac{kt_{nom}}{t_{max}} \tag{B.8}$$

Using the above equation and definitions, the c corresponding to the nominal process (c_{nom}) is given by the following expression:

$$c_{nom} = \frac{c_{high} + c_{low}}{2}. \tag{B.9}$$

Substituting c_{high} and c_{low} yields

$$c_{nom} = \frac{t_{nom}}{t_{max}}\left(\frac{2k-1}{2}\right). \tag{B.10}$$

Thus the amount of variation that can be tolerated about c_{nom} is

$$\text{range} = \frac{(c_{high} - c_{low})/2}{c_{nom}}. \tag{B.11}$$

Substituting c_{high}, c_{low}, and c_{nom} yields

$$\text{range} = \frac{1}{2k-1}. \tag{B.12}$$

REFERENCES

[1] S. Anderson, J. Earle, R. Goldschmidt, and D. Powers. The IBM system/360 model 91 floating point execution unit. *IBM Journal of Research and Development*, pages 34–53, January 1967.

[2] K. J. Antreich and M. H. Schulz. Accelerated Fault Simulation and Fault Grading in Combinational Circuits. *IEEE Transactions on Computer-Aided Design*, CAD-6(5):704–711, September 1987.

[3] R. L. Bauer, J. Fang, A. Ng, and R. K. Brayton. Simulating MOS VLSI circuits using SuperCrystal. In *VLSI 87*, 1987.

[4] R. L. Bauer, J. Fang, A. Ng, and R. K. Brayton. XPSim: a MOS VLSI circuit simulator. In *ICCAD-88*, 1988.

[5] M. Bazers and R. Ashuri. A novel CMOS digital clock and data decoder. *IEEE Journal of Solid State Circuits*, pages 1934–1940, December 1992.

[6] J. Benkoski. *Statistical Timing Verification and Delay Fault Detection by Formal Signal Interaction Modeling in a Multi-level Timing Simulator*. PhD thesis, Carnegie Mellon University, 1989.

[7] J. Benkoski, E. V. Meersch, L. Claesen, and H. De Man. Efficient algorithms for solving the false path problem in timing verification. In *ICCAD-87*, pages 44–47, 1987.

[8] J. Benkoski and A. J. Strojwas. Timing verification by formal signal interaction modeling in a multi-level timing simulator. In *26th Design Automation Conference*, pages 668–673, 1989.

[9] D. Brand and V. S. Iyengar. Timing analysis using functional analysis. In *ICCAD-86*, pages 126–129, 1986.

[10] R. P. Brent and H. T. Kung. A regular layout for parallel adders. *IEEE Transactions on Computer*, C-31(3), March 1982.

[11] R. E. Bryant. MOSsim: A switch-level simulator for MOS LSI. In *18th Design Automation Conference*, 1981.

[12] C.-H. Chang, E. S. Davidson, and K. A. Sakallah. Using constraint geometry to determine maximum rate pipeline clocking. In *ICCAD-92*, pages 142–148, 1992.

[13] T. I. Chappell, B. A. Chappell, S. E. Schuster, J. W. Allan, S. P. Klepner, R. V. Joshi, and R. L. Franch. A 2-ns cycle, 3.8 ns access 512-kb CMOS ECL SRAM with a fully pipelined architecture. *IEEE Journal of Solid State Circuits*, pages 1577–1585, November 1991.

[14] B. R. Chawla, H. K. Gummel, and P. Kozak. MOTIS - an MOS timing simulator. *IEEE Transactions on Circuits and Systems*, pages 901–910, December 1975.

[15] H. C. Chen and H. C. Du. Path sensitization in critical path problem. In *τ90: 1990 ACM Workshop on Timing Issues in the Specification and Synthesis of Digital Systems*, 1990.

[16] H. C. Chen and H. C. Du. Path sensitization in critical path problem. In *ICCAD-91*, pages 208–211, 1991.

[17] L. W. Cotten. Circuit implementation of high-speed pipeline systems. In *Proceedings of Fall Joint Computer Conference*, pages 489–504, 1965.

[18] L. W. Cotten. Maximum-rate pipeline systems. In *1969 AFIPS Proc. Spring Joint Computer Conference*, pages 581–586, 1969.

[19] S. Devadas, K. Keutzer, and S. Malik. Delay computation in combinational logic circuits: Theory and algorithms. In *ICCAD-91*, pages 176–179, 1991.

[20] S. Devadas, K. Keutzer, S. Malik, and A. Wang. Certified timing verification and transition delay of a logic circuit. In *29th Design Automation Conference*, pages 549–555, 1992.

[21] H. C. Du, H. C. Yen, and S. Ghanta. On the general false path problem in timing analysis. In *26th Design Automation Conference*, pages 555–560, 1989.

[22] B. Ekroot. *Optimization of Pipelined Processors by Insertion of Combinational Logic Delay.* PhD thesis, Stanford University, 1987.

[23] D. Erdman, D. Rose, and G. Nifong. CAzM: Circuit analyzer with macromodeling user's guide. Technical report, MCNC, June 1990.

[24] D. Fan, T. Gray, W. Farlow, T. Hughes, W. Liu, and R. Cavin. A CMOS parallel adder using wave pipelining. In *Advanced Research in VLSI and Parallel Systems*, Providence, RI, March 1992.

[25] B. Fawcett. Maximal clocking rates for pipelined digital systems. Master's thesis, University of Illinois, 1976. Technical Report R-706, Coordinated Science Laboratory.

[26] J. Fishburn. Clock skew optimization. *IEEE Transactions on Computers*, 39(7):945 – 951, 1990.

[27] P.D. Franzon, S. Simovich, M. Steer, M. Basel, S. Mehrotra, and T. Mills. Tools to aid in wiring rule generation for high speed interconnects. In *29th Design Automation Conference*, pages 466–471, 1992.

[28] S. Gaiotti, M. R. Dagenais, and N. C. Rumin. Worst-case delay estimation of transistor groups. In *26th Design Automation Conference*, pages 491–496, 1989.

[29] C. T. Gray, T. Hughes, D. Fan, G. Moyer, W. Liu, and R. Cavin. A high speed CMOS FIFO using wave pipelining. Technical Report NCSU-VLSI-91-01, North Carolina State University, 1991.

[30] T. Gray, T. Hughes, S. Arora, W. Liu, and R. Cavin. Theoretical and practical issues in CMOS wave pipelining. In *IFIP Transactions A: Computer Science and Technology - VLSI '91*, pages 397–409, Edinburgh, Scotland, 1992.

[31] A. B. Grebene. *Bipolar and MOS Analog Integrated Circuits*. Wiley, 1984.

[32] R. B. Hitchcock. Timing verification and the timing analysis program. In *19th Design Automation Conference*, 1982.

[33] M. Horowitz, A. Chan, J. Cobrunson, J. Gasbarro, T. Lee, W. Leung, W. Richardson, T. Thrush, and Y. Fujii. PLL design for a 500 MB/s interface. In *IEEE International Solid State Circuits Conference*, pages 160–161, February 1993.

[34] S. T. Huang, T. M. Parng, and J. M. Shyu. A new approach to solving false path problem in timing analysis. In *ICCAD-91*, pages 216–219, 1991.

[35] T. A. Hughes, C. T. Gray, W. Liu, and R. K. Cavin. Method and apparatus for high speed digital sampling of a data signal, March 25 1992. U.S. Patent Pending 07/857,457.

[36] N. Ishiura, M. Takahashi, and S. Yajima. Time-symbolic simulation for accurate timing verification of asynchronous behavior of logic circuits. In *26th Design Automation Conference*, pages 497–502, 1989.

[37] M. G. Johnson and E. L. Hudson. A variable delay line PLL for CPU-coprocessor synchronization. *IEEE Journal of Solid State Circuits*, pages 1218–1223, October 1988.

[38] N. P. Jouppi. TV: A nMOS timing analyzer. In *Proceedings Third Caltech VLSI Conference*, pages 71–85, 1983.

[39] D. A. Joy. *Clock Period Minimization with Wave Pipelining.* PhD thesis, University of Massachusetts, 1991.

[40] D. A. Joy and M. J. Ciesielski. Placement for clock period minimization with multiple wave propagation. In *28th Design Automation Conference*, pages 640–643, 1991.

[41] D. A. Joy and M. J. Ciesielski. Clock period minimization with wave pipelining. *IEEE Transactions on Computer-Aided Design*, pages 461–472, April 1993.

[42] S. T. Ju and C. W. Jen. A high speed multiplier design using wave pipelining technique. In *IEEE APCCAS*, pages 502–506, Australia, 1992.

[43] Y. C. Ju and R. A. Saleh. Incremental techniques for the identification of statically sensitizable critical paths. In *28th Design Automation Conference*, pages 541–546, 1991.

[44] Y. C. Ju and R. A. Saleh. Incremental circuit simulation using waveform relaxation. In *29th Design Automation Conference*, pages 8–11, 1992.

[45] B. Kim, D. N. Helman, and P. R. Gray. A 30-MHz hybrid analog/digital clock recovery circuit in 2 μm CMOS. *IEEE Journal of Solid State Circuits*, pages 1385–1394, December 1990.

[46] L. S. Kim and R. W. Dutton. Metastability of CMOS latch/flip-flop. *IEEE Journal of Solid State Circuits*, pages 942–951, August 1990.

[47] Y. H. Kim, S. H. Hwang, and A. R. Newton. Electrical-logic simulation and its application. *IEEE Transactions on Computer-Aided Design*, January 1989.

[48] T. W Kirkpatrick and N. Clark. PERT as an aid to logic design. *IBM Journal of Research and Development*, 1966.

[49] F. Klass. Personal communication.

[50] F. Klass and J. M. Mulder. CMOS implementation of wave pipelining. Technical Report 1-68340-44(1990)02, Delft University of Technology, December 1990.

[51] P. Kogge. *The Architecture of Pipelined Computers.* McGraw-Hill, 1981.

[52] Charlie Kohlhardt. PA-RISC processor for "Snake" workstations. In *Hot Chips Symposium*, pages 1.20–1.31, 1991.

[53] H. T. Kung and C. E. Leiserson. Systolic arrays (for VLSI). In *1978 Symposium on Sparse Matrix Computations and Their Applications*, pages 48–53, 1978.

[54] W. K. C. Lam, R. K. Brayton, and A. Sangiovanni-Vincentelli. Valid clocking in wavepipelined circuits. In *ICCAD-92*, Santa Clara, CA, November 1992.

[55] W. K. C. Lam, R. K. Brayton, and A. L. Sangiovanni-Vincentelli. Exact delay computation with timed boolean functions. Technical Report C92369, Semiconductor Research Corporation, July 1992.

[56] C. Leiserson and J. Saxe. Optimizing synchronous systems. *Journal of VLSI and Computer Systems*, 1(1):41–67, 1984.

[57] C. E. Leiserson, F. M. Rose, and J. B. Saxe. Optimizing synchronous circuitry by retiming. In *Proceedings of Third Caltech Conference on VLSI*, pages 87–116, 1983.

[58] W. Li, S. Reddy, and S. Sahni. Long and short covering edges in combination logic circuits. *IEEE Transactions on Computer-Aided Design*, 9(12):1245–1253, 1990.

[59] W. Lien and W. Burleson. Wave-domino logic: Theory and applications. Submitted to IEEE Transactions on Circuits and Systems.

[60] L. R. Liu, H. C. Chen, and H. C. Du. The calculation of signal stable ranges in combinational circuits. In *ICCAD-91*, pages 312–315, 1991.

[61] W. Liu, C. T. Gray, D. Fan, W. Farlow, T. Hughes, and R. K. Cavin. A 250 MHz wave pipelined adder in 2 μm CMOS. Technical Report NCSU-μSYS-92-03, North Carolina State University, 1992.

[62] H. Loomis. The maximum rate accumulator. *IEEE Transactions on Electronic Computers*, pages 628–639, August 1966.

[63] A. R. Martello, S. P. Levitan, and D. M. Chiarulli. Timing verification using HDTV. In *27th Design Automation Conference*, pages 118–123, 1990.

[64] S. R. Maskai, S. Kiaei, and D. J. Allstot. Synthesis techniques for CMOS folded source-coupled logic circuits. *IEEE Journal of Solid State Circuits*, pages 1157–1167, August 1992.

[65] R. Mayo, M. Arnold, W. Scott, D. Stark, and G. Hamachi. 1990 DECWRL/Livermore Magic release. Technical Report 90/7, WRL, September 1990.

[66] P. C. McGeer. *On the Interaction of Functional and Timing Behavior of Combinational Logic Circuits*. PhD thesis, University of California at Berkeley, 1989.

[67] P. C. McGeer and R. K. Brayton. Efficient algorithms for computing the longest viable path in a combinational network. In *26th Design Automation Conference*, pages 561–567, 1989.

[68] P. C. McGeer and R. K. Brayton. *Integrating Functional and Temporal Domains in Logic Design*. Kluwer Academic Publishers, 1991.

[69] P. C. McGeer, A. Saldanha, P. R. Stephan, R. K. Brayton, and A. L. Sangiovanni-Vincentelli. Timing analysis and delay-fault test generation using path-recursive functions. In *ICCAD-91*, pages 180–183, 1991.

[70] P. C. McGeer, A. Saldanha, P. R. Stephan, R. K. Brayton, and A. L. Sangiovanni-Vincentelli. Delay models and sensitization criteria in the false path problem. Technical Report C92555, Semiconductor Research Corporation, October 1992.

[71] Meta-Software, Inc., Campbell, CA. *HSPICE User's Manual.*

[72] L. W. Nagel. SPICE 2: A computer program to simulate semiconductor circuits. Technical Report ERL-M510, Electronics Research Laboratory, University of California, Berkeley, 1975.

[73] V. Nguyen, W. Liu, C. T. Gray, and R. K. Cavin. A CMOS multiplier using wave pipelining. In *Custom Integrated Circuits Conference*, pages 12.3.1–12.3.4, San Diego, CA, May 1993.

[74] V. D. Nguyen. A CMOS multiplier using wave pipelining. Master's thesis, North Carolina State University, 1993.

[75] J. K. Ousterhout. Crystal: A timing analyzer for nMOS VLSI circuits. In *Proceedings Third Caltech VLSI Conference*, pages 57–69, 1983.

[76] S. Perremans, L Claesen, and H. De Man. Static timing analysis of dynamically sensitizable paths. In *26th Design Automation Conference*, pages 568–573, 1989.

[77] L. Pillage. *Asymptotic Waveform Evaluation for Timing Analysis*. PhD thesis, Carnegie Mellon University, 1989. SRC Report CMUCAD-89-34.

[78] L. Qi and X. Peisu. The design and implementation of a very fast experimental pipelining computer. *Journal of Computer Science and Technology*, 3(1):1–6, 1988.

[79] T. Quarles. *The SPICE3 Circuit Simulator*. PhD thesis, University of California at Berkeley, 1989.

[80] T. Rahkonen, J. Kostamovaara, and S. Saynajakangas. Time interval measurement using integrated tapped CMOS delay lines. In *32nd Midwest Symposium on Circuits and Systems*, pages 201–205, 1990.

[81] J. P. Roth. Diagnosis of automata failures: A calculus and a new method. *IBM Journal of Research and Development*, pages 278–281, 1966.

[82] R. Rudell and A. Sangiovanni-Vincentelli. Multiple-valued minimization for PLA optimization. *IEEE Transactions on Computer-Aided Design*, 6(5):727–750, September 1987.

[83] K. Sakallah, T. Mudge, T. Burks, and E. Davidson. Synchronization of pipelines. Technical Report CSE-TR-97-91, Computer Science and Engineering, University of Michigan, February 1991.

[84] K. Sakallah, T. Mudge, and O. Olukotun. A timing model of synchronous digital circuits. Technical Report CSE-TR-47-90, Computer Science and Engineering, University of Michigan, March 1990.

[85] T. Sakurai. Optimization of CMOS arbiter and synchronizer circuits with submicrometer MOSFET's. *IEEE Journal of Solid State Circuits*, pages 901–906, August 1988.

[86] T. Sakurai and A. R. Newton. Alpha-power law MOSFET model and its application to CMOS inverter delay and other formulas. *IEEE Journal of Solid State Circuits*, pages 584–594, April 1990.

[87] T. Sakurai and A. R. Newton. Delay analysis of series-connected MOSFET circuits. *IEEE Journal of Solid State Circuits*, pages 122–131, February 1991.

[88] A. Salz and M. Horowitz. IRSIM: An incremental mos switch-level simulator. In *26th Design Automation Conference*, pages 173–178, 1989.

[89] Y. H. Shih and S. M. Kang. ILLIADS: A new fast MOS timing simulator using direct equation-solving approach. In *28th Design Automation Conference*, pages 20–25, 1991.

[90] H. Shin, J. Warnock, M. Immediato, K. Chin, C.-T. Chuang, M. Cribb, D. Heidel, Y.-C. Sun, N. Mazzeo, and S. Brodskyi. A 5Gb/s 16x16 Si-bipolar crosspoint switch. In *IEEE International Solid State Circuits Conference*, pages 228–229, February 1992.

[91] J. P. Silva, K. A. Sakallah, and L. M. Vidigal. FPD - an environment for exact timing analysis. In *ICCAD-91*, pages 212–215, 1991.

[92] A. Srinivasan and D. LaPotin. Cycle time optimization subject to packaging constraints. In *Multichip Module Workshop*, Santa Cruz, CA, March 1991.

[93] A. Srinivasan, S. Seshadri, J. G. Shanthikumar, and D. P. LaPotin. Stochastic cycle time optimization of sequential systems. Technical report, University of California, Berkeley, Electronics Research Laboratory, December 1990. Technical Memo UCB/ERL M90/91.

[94] A. D. Stein, T. V. Nguyen, B. J. George, and R. A. Rohrer. ADAPTS: A digital transient simulation strategy for integrated circuits. In *28th Design Automation Conference*, pages 26–31, 1991.

[95] R. Stewart and J. Benkoski. Static timing analysis using interval constraints. In *ICCAD-91*, pages 308–311, 1991.

[96] H. Stone. *High-Performance Computer Architecture.* Addison Wesley, 1987.

[97] T. G. Szymanski. LEADOUT: A static timing analyzer for MOS circuits. In *ICCAD-86*, pages 130–133, 1986.

[98] C. J. Termnan. RSIM - a logic-level timing simulator. In *Proceedings of ICCD-83*, 1983.

[99] S. Unger and C. Tan. Clocking schems for high-speed digital systems. *IEEE Transactions on Computers*, 35(10):880 – 895, 1986.

[100] W. van Noije, C. T. Gray, W. Liu, T. A. Hughes, R. K. Cavin, and W. J. Farlow. CMOS sampler with 1 gbit/s bandwidth and 25 ps resolution. In *Custom Integrated Circuits Conference*, pages 27.5.1–27.5.4, San Diego, CA, May 1993.

[101] W. van Noije, W. Liu, and J. Navarro S. Jr. Metastability behavior of mismatches CMOS flip-flops using state diagram analysis. In *Custom Integrated Circuits Conference*, pages 27.7.1–27.7.4, San Diego, CA, May 1993.

[102] W. A. M. van Noije, C. T. Gray, W. Liu, and R. K. Cavin. CMOS sampling circuitry with 25 ps resolution. Technical Report NCSU-μSYS-92-05, North Carolina State University, 1992.

[103] L. M. Vidiagal, S. R. Nassif, and S. W. Director. CINAMON: Coupled integration and nodal analysis of MOS networks. In *23th Design Automation Conference*, pages 179–185, 1986.

[104] C. Visweswariah and R. A. Rohrer. SPECS2: An integrated circuit timing simulator. In *ICCAD-87*, pages 94–97, 1987.

[105] J. White and A. L. Sangiovanni-Vincentelli. *Relaxation-Based Circuit Simulation.* Kluwer Academic Publishers, 1985.

[106] T. W. Williams, B. Underwood, and M. R. Mercer. The interdependence between delay-optimization of synthesized networks and testing. In *28th Design Automation Conference*, pages 87–92, 1991.

[107] D. Wong. *Techniques for Designing High-Performance Digital Circuits using Wave Pipelining.* PhD thesis, Stanford University, September 1991.

[108] D. Wong, G. DeMicheli, and M. Flynn. Inserting active delay elements to achieve wave pipelining. Technical Report CSL-TR-89-386, Computer System Laboratory, Stanford University, 1989.

[109] D. Wong, G. De Micheli, and M. Flynn. Designing high-performance digital circuits using wave pipelining. In *VLSI '89*, Munich, W. Germany, August 1989.

[110] D. C. Wong, G. De Micheli, and M. J. Flynn. Designing high-performance digital circuits using wave pipelining: Algorithms and practical experiences. *IEEE Transactions on Computer Aided Design*, pages 25–46, January 1993.

[111] D. C. Wong, G. De Micheli, M. J. Flynn, and R. E. Huston. A bipolar population counter using wave pipelining to achieve 2.5x normal clock frequency. *IEEE Journal of Solid State Circuits*, pages 745–753, May 1992.

[112] H. C. Yen, H. C. Du, and S. Ghanta. Efficient algorithms for extracting the K most critical paths in timing analysis. In *26th Design Automation Conference*, pages 649–654, 1989.

[113] H. C. Yen, S. Ghanta, and H. C. Du. A path selection algorithm for timing analysis. In *25th Design Automation Conference*, pages 720–723, 1988.

[114] I. A. Young, J. K. Greason, and K. L. Wong. A PLL clock generator with 5 to 110 MHz of lock range for microprocessors. *IEEE Journal of Solid State Circuits*, pages 1599–1607, November 1992.

[115] H. Youssef, E. Shragowitz, and L. C. Bening. Critical path issue in VLSI design. In *ICCAD-89*, pages 520–523, 1989.

[116] C. Zukowski and P. S. Lin. Differential high-bandwidth communication circuits. In *Sixth MIT Conference on Advanced Research in VLSI*, pages 279–297, 1990.

[117] C. Zukowski and K. Shum. A matched-delay CMOS TDM multiplexer cell. In *IEEE ICCD*, pages 352–355, 1988.

INDEX